Making Use of Deleuze i

Making Use of Deleuze in Planning translates and re-creates some of Gilles Deleuze's most abstract philosophical concepts to form a new, practicable planning assessment tool. It shows what his philosophy can do for planning theory as well as planning assessment practice and, in doing so, sets out a pragmatic approach to Deleuzian studies: one that helps form bridges between ontological problems and the problems found in professional practice. It also breaks new ground in assessment methodology by challenging the essentialist ideas underpinning assessment methods like BREEAM and setting out and testing a new form of non-essentialist assessment named SIAM. The book argues that Deleuze's philosophy can be made useful to planning as long as one is prepared to adapt and re-create his key ontological concepts to respond to the specific demands of the field.

Gareth Abrahams crosses an invisible line between theory and practice. He is a Deleuzian theorist whose publications push the boundaries of Deleuzian scholarship and spatial theory. He is also a practising, chartered architect who has designed and delivered a range of complex buildings across the UK from specialist schools to large office developments and from bespoke new-build schemes to highly sensitive heritage refurbishments.

New Directions in Planning Theory

Series Editors

Gert de Roo, University of Groningen, The Netherlands
Jean Hillier, RMIT University, Australia
Joris Van Wezemael, University of Fribourg, Switzerland

Routledge's series, New Directions in Planning Theory, develops and disseminates theories and conceptual understandings of spatial and physical planning which address such challenges as uncertainty, diversity and incommensurability.

Planning theories range across a wide spectrum, from questions of explanation and understanding, to normative or predictive questions of how planners should act and what future places should look like.

These theories include procedural theories of planning. While these have traditionally been dominated by ideas about rationality, in addition to this, the series opens up to other perspectives and also welcomes theoretical contributions on substantive aspects of planning.

Other theories to be included in the series may be concerned with questions of epistemology or ontology; with issues of knowledge, power, politics, subjectivation; with social and/or environmental justice; with issues of morals and ethics.

Planning theories have been, and continue to be, influenced by other intellectual fields, which often imbue planning theories with awareness of and sensitivity to the multiple dimensions of planning practices. The series editors particularly encourage inter- and trans-disciplinary ideas and conceptualisations.

Making Use of Deleuze in Planning

Proposals for a speculative and immanent assessment method

Gareth Abrahams

Routledge
Taylor & Francis Group
LONDON AND NEW YORK

First published 2017 by Routledge

2 Park Square, Milton Park, Abingdon, Oxfordshire OX14 4RN
711 Third Avenue, New York, NY 10017

Routledge is an imprint of the Taylor & Francis Group, an informa business

First issued in paperback 2018

British Library Cataloguing-in-Publication Data
A catalogue record for this book is available from the British Library

Library of Congress Cataloging-in-Publication Data
Names: Abrahams, Gareth, author.
Title: Making use of Deleuze in planning : proposals for a speculative and immanent assessment method / Gareth Abrahams.
Other titles: Making Deleuze useful to planning
Description: New York : Routledge, 2017. | Series: New directions in planning theory | Revision of the author's thesis (Cardiff University) under the title: Making Deleuze useful to planning. | Includes bibliographical references and index.
Identifiers: LCCN 2016013660| ISBN 9781472477576 (hb : alk. paper) | ISBN 9781315593395 (ebook)
Subjects: LCSH: Architecture—Planning. | Deleuze, Gilles, 1925-1995.
Classification: LCC NA2500 .A395 2017 | DDC 720/.1—dc23
LC record available at https://lccn.loc.gov/2016013660

ISBN: 978-1-4724-7757-6 (hbk)
ISBN: 978-1-138-39280-9 (pbk)

Typeset in Times New Roman
by Apex CoVantage, LLC

Contents

Figures

Tables

Notes on author

I began my architecture degree in 1998 and was immediately captured by the idea that architectural decisions should be driven by a strong ontological position rather than aesthetic opinion and rules of thumb. In 2001 I studied at the Ecole d'architecture, urbanisme and paysage in Lille, where I was introduced to the philosophical texts of Gilles Deleuze. Three years later I qualified as a chartered architect, and, over the following twelve years, I went on to design and run many schemes from inception to completion.

But across this period I kept these two worlds apart: practising architect in the day and Deleuzian scholar at night. My academic interest is now directed at bringing these two worlds together to answer a question that has troubled me since 1998 and continues to trouble so many architectural students and practitioners alike: once I have developed a strong theoretical line of enquiry, how do I make the transition to a design, a set of robust details and a physical building? In other words, how does one make philosophy useful to practitioners sat at their desks and drawing boards? Answering this question forms the basis of this book and the basis of my academic agenda in the years to come.

Preface and acknowledgements

I dedicate this book to my beloved wife, Sarah (Kitty), who I have bombarded with my Deleuzian rantings for more years than I should.

I would like to thank Taylor and Francis for the opportunity to publish this book, Dr Neil Harris and Dr Richard Cowell for their enthusiasm during my PhD and Professor Jean Hillier for our on-going Deleuzian conversation which has been a source of constant stimulation.

Introduction

A pragmatic approach to Deleuze the pragmatist

The philosopher Gilles Deleuze and the psychologist Felix Guattari introduce one of their most cited texts *A Thousand Plateaus* with the phrase:

> The two of us wrote Anti-Oedipus together. Since each of us was several, there was already quite a crowd.
>
> (Deleuze and Guattari, 2004b: 3)

In many ways I start this thesis with the same feeling: the same 'schiz' between different authors. Whilst this is a single-authored monograph, I feel as though I am writing it in co-authorship: Gareth as academic planning theorist in collaboration with Gareth as practicing architect; Gareth as abstract thinker with Gareth as pragmatist. It is between these different authors that I set out my objective for this book, which is to consider whether Deleuze's ontological concepts can be of practicable use not just to the analyst operating on the edge of professional practice but also to the planner, masterplanner and architect sat at their desks and drawings boards.

This pragmatic approach to Deleuze sets out different questions to the ones often captured in much of the literature based on Deleuze's philosophy. Many of us drawn to Deleuzian concepts will have come across and digested Deleuzian dictionaries (Parr, 2005), reconstructions of Deleuze's ontological framework (DeLanda, 2002) or expanded ontological studies based on one or several Deleuzian concepts (see DeLanda's assemblage theory [2006] and Bryant's machinic ontology [2014]). These texts are invaluable resources and help us understand and contextualize concepts like the assemblage, machines, becomings, mulitiplicities and the plane of immanence. But they are mostly focused on a 'what is' line of enquiry: '*what is* an assemblage, a becoming or multiplicities?' or '*what is* a Deleuzian or Deleuze-inspired ontology?' The questions that interest me are far more pragmatic in nature: 'what does the assemblage, becomings and multiplicities *do*?' and more importantly still, 'what might these concepts *do for me*?'

These pragmatic questions are not in conflict with Deleuze's own line of enquiry. As the planning theorist Jean Hillier notes, Deleuze's method is probably best described as a form of

> generative pragmatics where practice is radical experimentation; a power-full, collaborative 'politics of language' – a sort-of 'pragmatism plus'!
>
> (Hillier, 2011: 504)

With this in mind, I believe that this study is best described as a pragmatic approach to Deleuze the pragmatist and, as such, appeals to theorists and practitioners who are equally struggling to unite the different voices in their heads: their own schiz.

Objective of this book

My objective for writing this book, therefore, is to explore whether Deleuze's philosophical concepts *can* be usefully translated into the 'planning project' (Healey, 2010), and, if so, to identify the measures needed to create a Deleuze-inspired planning tool for use in planning/development practice. In doing so, my aims broadly reflect Patsy Healey's efforts to

> extract . . . the planning project from the narrowing, reductive perspective with which its many practices have become associated in the later twentieth century [whilst considering] . . . how ideas . . . get transformed into programmes of action, which then have material effects on living conditions and local environments.
>
> (Healey, 2010: ix, x)

Like Healey, I hope that this approach to the 'planning project' can go beyond attempts to 'imagine futures' and find ways to 'bring . . . futures into being'. (Healey, 2010: x)

Why Deleuze?

The most obvious starting point for tackling a study such as this is to explain *why* I have decided to focus on Deleuze's philosophy.

Why Deleuze and planning?

Cross-disciplinary by 'design'

The first point I would like to make is taken from Deleuze's philosophy itself and his insistence that applying concepts across different bodies of knowledge was integral to his own method. In his interview with Clare Parnet, Deleuze notes,

> It is not the elements or the sets which define the multiplicity.[1] What defines it is the AND, as something which has its place between the elements or between the sets. AND, AND, AND – stammering.
>
> (Deleuze and Parnet, 2002: 26)

In this quotation, Deleuze argues that concepts should not be created and used within one field alone. Rather, concepts created in one body of knowledge, such as philosophy, should be added to concepts found in other bodies of knowledge. Deleuze goes on to suggest that the relationships formed by such 'encounters' offer new directions for development that do not belong to either field but 'work between the two' (Deleuze and Parnet, 2002: 13). This additive 'AND, AND, AND' process can be seen across all of Deleuze's seminal texts. One of Deleuze and Guattari's most cited works, *A Thousand Plateaus*, for example, is formed as a series of chapters, or plateaus (Deleuze and Guattari, 2004b). Each plateau reveals an encounter between Deleuze and Guattari's philosophical concepts and the concepts created and used within a specific field, or milieu, such as linguistics, archaeology, geology and psychology. Thus, it seems that Deleuze created his concepts by drawing on links with other fields of knowledge and intended these concepts to be used in the same way. In other words, Deleuze's concepts are 'designed' to be used both outside, and within, philosophy.

A growing trend towards Deleuzian philosophy

A quick Google search reveals that this intellectual agenda of applying Deleuze to other disciplinary spheres, including planning, has already been growing apace. This trend has been actively encouraged by key figures in the Deleuzian community. The *Deleuze Connections* series, for example, uses the above 'AND, AND, AND' quotation to introduce each of its publications. Commenting on this approach, the series editor, Ian Buchanan, notes that such studies have

> already placed Deleuze's thought in connection with feminist theory, music, space, geography, queer theory, performance, postcolonial studies, contemporary art, and [are] constantly opening new frontiers in Deleuze Studies.
>
> (Buchanan, no date)

The list of fields noted by Buchanan demonstrates how broad these encounters have become. Several studies in the series, such as Martin-Jones and Brown's study of film (2012), build on conceptual relationships already explored within Deleuze's seminal texts, whilst others, such as Frichot and Loo's study of architecture (2013) and Frichot et al.'s study of the city (2016) build on less prominent fields in his philosophy.

This cross-disciplinary approach is not limited to this specific community of Deleuzian academics. In 2002, the esteemed Deleuzian scholar Manuel DeLanda outlined Deleuze's ontology from within the sciences (2002) and went on to develop a 'neo-Deleuzian' assemblage theory by working within a range of spatial disciplines (2006), and spatial planning in particular (2010). In 2007, the planning theorist Jean Hillier published a detailed proposal for a new Deleuze-inspired approach to strategic planning (Hillier, 2007). Several years later, Hillier's co-edited book on planning theory included a further development of these ideas (Hillier, 2011) alongside several other articles that used Deleuze's concepts to critically engage with issues particular to the field (Pløger, 2010b; Van Wezemael, 2010). These efforts have been expanded by other more recent articles in planning theory (Purcel, 2013; Ansaloni and Tedeschi, 2015 for example). This study, therefore, is supported by this growing trend seen in a range of disciplines including planning theory.

Gaps in knowledge

My third reason for focusing on Deleuze's philosophy relates to the gaps left unanswered in this growing trend towards, what one might term, a *Deleuze-inspired* approach to planning. One such gap can be seen in the seemingly obscure and abstract terminology surrounding Deleuzian philosophy. If we look through Deleuze's philosophical texts, it is difficult to imagine how his concepts could relate to debates in planning literature, and even more difficult to link these concepts to the day-to-day work of the planner, masterplanner and architect sat at their desk or drawing board.

Whilst some efforts have been made to improve these links (Bonta and Protevi, 2004; Parr, 2005; Hillier, 2007), many of the definitions offered for concepts like 'assemblages', 'becomings' and 'multiplicities' draw on other parts of Deleuze's philosophy and do not acknowledge or explain why a range of meanings can be found in different texts. This problem stems, in part, from the fact that Deleuze's concepts are used to discuss very abstract ideas and problems. However, one might also conclude that this reflects an oversight amongst this growing community of planning theorists that prevents others from linking Deleuze's philosophical ideas to 'real'[2] situations.

Building on this final point, another gap in the literature surrounds questions of usefulness. Such questions include the following: what are the situations and problems for which Deleuze's concepts can be made useful? Can Deleuze's philosophy be translated into planning theory and practice or just the former? What is required to make these transitions? Taken together, these questions respond to a much broader issue about *what* can be reasonably expected from planning's engagement with Deleuze's philosophy as well as *when* and *where* these expectations can be met. Answering these kinds of questions would help planning theorists to speculate about the future of this growing trend in Deleuze-inspired planning and whether or not it can be used to meet the aspirations of some of its key figures – namely, to draw on Deleuze's concepts to create '. . . new practices and tools to increase the effectiveness of [the discipline]' (de Roo et al., 2012: 20).

To date, these aspirations have gone largely unfulfilled. One explanation for this can be seen in a continued reluctance to adapt and expand these concepts to meet the specific demands of engagement rather than analysis and critique (see Chapter 3). Probably the most successful and convincing attempt can be seen in Hillier's proposals for a new 'multiplanar' approach to strategic planning (2007; 2011). However, as I will show in Chapter 3, these attempts remain tentative and do not, as yet, show how Deleuze's concepts can be used to engage with, rather than analyse, existing forms of planning practice.

The concerns I raise above reflect a much broader on-going concern in the discipline, as expressed by the planning theorist John Forester. For Forester, the weighting towards Deleuze-inspired planning theory rather than Deleuze-inspired planning practice reflects a historical pattern in the field. In his critique of planning theory, Forester argues that most planning theorists use philosophical concepts to 'deconstruct' existing tools and practices but very rarely develop these critiques further by 'reconstructing' new methods of engagement (Forester, 2007).

This critique reflects a further gap in the literature surrounding the use of Deleuze's concepts. Planning theorists are yet to offer a thorough critical engagement with the relationship between Deleuze and planning. Whilst previous studies questioned the use of post-structuralism and post-modernism in planning and geography (Martin, 2001; Hamnett, 2003), these efforts failed to appreciate Deleuze's unique position in this broad group. As DeLanda notes,

> When confronted with Deleuze's original texts this audience is bound to be puzzled, and may even be repelled by the superficial similarity of these texts with books belonging to what has come to be known as the 'post-modern' tradition. Although . . . Deleuze has absolutely nothing in common with that tradition, his experimental style is bound to create that impression.
>
> (DeLanda, 2002: 1)

If we look at this relationship from another angle, we see there have been equally few efforts made to critically review, or understand, proposals made to link Deleuze's philosophy to planning processes and problems. Nyseth et al. (2009) provide the only notable attempt to review Hillier's proposal in the context of planning. This review successfully connects Hillier's Deleuze-inspired framework with existing processes of strategic planning but fails to explain *how* Hillier makes the transition from philosophy to planning theory and practice, what adaptations were made, what gaps and questions were left unanswered, and what this tells us about the viability of making Deleuze's philosophy useful to planning practice.

These three gaps in the literature surrounding Deleuze and planning provide the basis for, and the principle contributions of, this text. I make these contributions by:

1. providing a clearer understanding of Deleuze's key philosophical concepts. This is achieved by relating them to 'real' situations and offering guidance about how one might use these concepts in subsequent studies;
2. exploring possibilities for translating Deleuze's concepts from philosophy to planning theory and planning practice, and showing what is needed to

operationalise these concepts in order to create 'new practices and tools to increase the effectiveness of [the discipline]' (de Roo et al., 2012: 20);

3 critically engaging with other Deleuze-inspired studies and developing my own Deleuze-inspired study to offer subsequent theorists a realistic expectation about *what* can be achieved by this Deleuze/planning link, as well as *where* and *when* this link might prove useful and effective.

What is the specific problem, and why might Deleuze help to resolve this problem?

Above I set out a broad argument for looking at Deleuze's philosophy in the field of planning and show how this argument is used to direct this book as a contribution to the literature. In my second response to the question 'why Deleuze', I will briefly show how Deleuze relates to a specific problem within this field.

A problem

This specific problem focuses on the ideas used to form certain kinds of assessment tools as a key part of the broader 'planning project'. This area of planning has been the focus for a number of pragmatists in planning theory who have argued against the use of simple, essentialist definitions as 'first principles'(Healey, 2009b). Such 'essentialist' definitions reflect the idea that concepts can be defined according to their inherent features or traits (see Chapter 2). The results of such principles have been well documented by this, and other, areas of the planning community. Planning tools derived from these essentialist definitions are thought to be overly rigid and unable to account for the subtle differences in meaning held by different parties or reflecting different contextual features. Such rigidity has, in several documented instances, exacerbated conflicts during planning discussions (Stein and Harper, 2012), has marginalised certain interest groups within such discussions (Richardson and Jensen, 2001; 2003: 19) and has concluded in actions that could not be implemented in specific contexts (Jamal et al., 2002). These problems are reflected in the experiences of other actors working in the design and development process (including my own) (Schweber, 2013) and form the basis for proposed changes to the use of regulation in the United Kingdom (DCLG, 2013; 2014).

These calls for a new approach to the way planning tools are conceived are equally captured in planning assessment literature (Borri, 1998). As in the proposals made by pragmatists like Stein and Harper (2012), these calls point to a flexible, context-sensitive form of assessment designed to accommodate the differing and changing values of actors working on a specific scheme. To date, these calls have gone mostly unrealised (Alexander, 2006). A number of explanations and arguments have been made to explain (Baum, 1996; Borri, 1998; Lichfield and Prat, 1998) or justify (Fischer, 2003) this resistance to such contextually sensitive assessments and to highlight the important roles played by essentialist assessment – namely, their links to broader policy objectives and their suitability to cross-site comparison. Such efforts do not, however, satisfy the continuing demand placed on assessment methodologists to meet these calls (Richardson, 2005; Alexander, 2006). This debate is reviewed in Chapter 2.

Why Deleuze's philosophy might help contribute to a new non-essentialist form of assessment

The search for solutions within planning theory and philosophy are reflected in Richardson's calls for new ways of 'seeing' and developing assessment methods (Lawson, 2006; Richardson, 2005). Deleuze's relevance to, what one might broadly term, the essentialist problem can be found in DeLanda's detailed review and reconstruction of Deleuze's philosophy. In this study DeLanda argues that many of Deleuze's concepts are conceived as alternatives to essentialism. Concepts like 'the assemblage', 'becomings' and 'multiplicities' are, he demonstrates, the result of non-essentialist lines of argument. These concepts, therefore, offer new opportunities for tackling the dominant role played by essentialist principles in assessment design and the problems that result from the rigidity of an essentialist definition unable to account for different meanings and the unique issues and opportunities of a given context.

By exploring this essentialist problem in assessment design, and considering how Deleuzian concepts might help direct new non-essentialist tools, this book makes a second set of contributions to the literature by:

1 offering new conceptual frameworks for distinguishing between and discussing essentialist and non-essentialist assessment methods;
2 offering new insight into the role played by these different kinds of assessment, as well as the role played by judgement in planning.

Two questions

The line of enquiry outlined above can be summarised as two questions that will provide a focus for subsequent chapters in this book:

1 Can Deleuze's philosophical concepts be translated into a new theoretical framework for constructing formal assessments?
2 If so, can this be translated into a practical tool useful to actors in the planning and development process?

Structuring a response to the research questions

As noted above, the principle objective of this book is to consider whether Deleuze's concepts can be translated from philosophy into planning theory and into a practicable planning tool. This broad objective points to established research practice, in which one moves from the abstract/conceptual/theoretical to the detailed/practical/empirical. However, Deleuze and Guattari argue that such transitions are best achieved by starting from the specific and moving *between* the abstract and the practical. In their introduction to the French edition of *A Thousand Plateaus*, they note that one should

> partir au milieu, par le milieu, entrer et sortir, non pas commencer ni finir. . . . Ils ont su faire une pragmatique. C'est que le milieu n'est pas du tout une moyenne, c'est au contraire l'endroit où les choses prennent de la vitesse.[3]
>
> (Deleuze and Guattari, 1980: 37)

In this quotation, Deleuze and Guattari stress the importance of starting from within a specific field of knowledge or situation (a milieu) and then developing the study by moving in and out of the field: from specific concerns, relationships and debates identified in the field, to concerns, relationships and debates in associated fields. It is only by starting from a milieu and keeping it in the forefront of one's mind, they argue, that ideas can be developed quickly and effectively. This structure provided me with a general methodology for undertaking and organising this study into Deleuze-inspired planning, as shown in the following breakdown of the chapters.

Part 1: Assessments, essentialism and Deleuze

As per Deleuze and Guattari's quote above, I introduce the kernel of a problem by starting from within a specific milieu identified from my nine years of professional experience working as a qualified architect. I use this milieu to outline a problem related to existing sustainable building assessments and expand this into broader literature. As Deleuze and Guattari note, I keep this milieu at the forefront of my mind and return to it in different instances to make sense of abstract, theoretical concepts (Chapter 2), to form comparisons with the way others have used these theoretical concepts (Chapter 3) or to work through a developing theoretical framework (Chapter 6). As per Deleuze and Guattari's comment, it was in such instances that I felt that the study gained the greatest clarity, which, in turn, helped me develop and test ideas efficiently and effectively.

In the second chapter, I expand this problem further. By relating it to literature from within planning theory, I show that the problem results from the essentialist principles used to construct and operationalise a definition of 'the sustainable home'. I develop this second chapter by reviewing some of the key challenges to essentialism within planning literature. The limitations I identify in this literature provide a framework for introducing and framing the concepts that form Deleuze's ontology.

To establish this link between planning theory and Deleuzian philosophy, I was keen to avoid the dictionary definition format seen in a number of other Deleuze-inspired studies, such as those discussed in Chapter 3 (Bonta and Protevi, 2004; Halsey, 2006; Hillier, 2007). Whilst dictionary definitions provide a useful reference point for the reader, they suggest that such concepts have a clear definition outside of their use. Yet, as noted earlier in this introduction, Deleuze argues that concepts alone do not define a framework. Rather, such concepts have meaning only when they assume their place 'between' the elements that form a particular field (Deleuze and Parnet, 2002: 26). With this in mind, I define Deleuze's key ontological concepts within the flow of the text, illustrated, as Deleuze and Guattari suggest, through constant shifts between the abstract ideas presented in their work and the specific milieu discussed in Chapter 1.

Taken together, these two chapters demonstrate the breadth of a problem that spans between philosophy, different areas of planning theory and planning practice.

Part 2: How to make Deleuze useful

In the third chapter, I turn to the issue of translating Deleuze's philosophy into a theoretical framework applicable to a given line of enquiry. This is a key aspect of the two research questions, which I explore by drawing lessons from a detailed review of three Deleuze-inspired studies positioned within the spatial disciplines. Unlike 'traditional' literature reviews, this approach allows for a much more detailed understanding of the relationship formed between a specific problem and Deleuze's philosophical concepts. In doing so, I am able to explain *how* these studies make the transition from philosophy to their respective field of interest, what adaptations were made, what gaps and questions were left unanswered, and what this tells us about the viability of making Deleuze's philosophy useful to spatial problems. In doing so, this approach helps respond to one of the gaps noted above.

My review of these Deleuze-inspired studies suggests that, to make Deleuze's ontological concepts useful, one cannot rely on the definitions inferred from Deleuze's texts alone. Doing so leads to significant gaps and unanswered questions in the resulting theoretical framework. Rather, I show that Deleuze's concepts must be adapted, re-created and expanded to respond to the demands of a specific field whilst retaining a broader sense of their ontological role in Deleuze's philosophy.

Part 3: A case study of Building Research Establishment (BRE) assessments

In the five chapters that form Part 3, I 'return' to the milieu identified in Chapter 1. Drawing on the broader, theoretical lessons learned from the preceding two chapters, I ask whether Deleuze's concepts can be translated into a theoretical proposal for a new assessment tool, and whether there is any empirical evidence to suggest that this tool could be operationalised in practice.

Chapter 4 explains why I have decided to focus on BRE assessments over other essentialist tools used in planning practice such as Environmental Impact Assessments and European Territorial Cohesion Indicators, and it identifies a research strategy for responding to the two research questions above.

Chapters 5 and 6 form the two parts to Research Stage A. In the first of these chapters, I develop a pragmatic method for developing a new Deleuze-inspired assessment tool. In the second chapter in this stage (Chapter 6), I outline the experiments I have undertaken to develop my proposals for SIAM: a Speculative and Immanent Assessment Method. This proposal takes its name from the principles used in its construction. It is an assessment method undertaken by key members of the design and development team; a method based on speculations about what might become of the project rather than the resulting properties of a completed design (Speculative); and a method undertaken as part of, and during, the design and development process rather than at pre-conceived stages of development (Immanent).

Chapters 7 and 8 form the two parts to Research Stage B intended to test this proposed tool through empirical enquiry. In the first of these chapters, I develop an empirical method in-keeping with Deleuze's stance on transcendent empiricism. Drawing on interviews and subsequent feedback from a sample of professionals working in design and regulatory roles, Chapter 8 outlines the viability of the proposed SIAM method.

Taken together, these four chapters show how I have adapted, re-created and expanded Deleuze's concepts to form a theoretical tool for planning practice, and they consider what further changes might be needed to operationalise this tool.

Part 4: Synthesis, discussion and conclusions

In Chapter 9 I return to the two research questions noted above and consider what the study into sustainable assessments tells us about making Deleuze useful to planning theory and practice. More broadly, I explore what other issues are revealed through these questions: issues relating to the way we think about the link between theory and practice, as well as the facilitating and regulatory roles of planning.

In the concluding chapter, Chapter 10, I bring together the points raised across the book as a whole to identify and review the contributions made by this study. As noted above, these include:

1 providing a clearer understanding of Deleuze's key philosophical concepts;
2 showing what is needed to operationalise Deleuze's concepts in planning practice;
3 critically engaging with other Deleuze-inspired studies;
4 offering new conceptual frameworks for assessment methods;
5 offering new insight into the role played by assessments and planning.

This final chapter goes on to identify the study's limitations, as well as opportunities to overcome these limitations in future research. In doing so, I hope to provide some direction for those working in Deleuzian scholarship, planning theory and planning practice, and, more importantly, for a small and growing community interested in translating Deleuze's concepts across all three of these fields.

Notes

1 The concept of the 'multiplicity' is explained and discussed in Chapter 3.
2 The use of the term 'real' in this sense should be understood as the common usage of the term and should be distinguished from the ontological use of the term, discussed later in the book.
3 I translate this as follows: 'Start from a specific field, by that field, enter and leave it without starting or finishing. . . . This is how one acts pragmatically. This field [also translates as middle] should not be understood as an average. It is quite the contrary, for it is in this place that things achieve the greatest speed'. Please note that my decision to draw on the original, French text rather than the English translation is based on the definition of the term 'milieu' (Deleuze and Guattari, 2004b). As my translation notes, the French use of this term has two meanings, which were not fully captured in the English translation, yet are important to the points raised in this chapter.

Part 1

Assessments, essentialism and Deleuze

Deleuze argued that one should start a study by placing oneself within a specific milieu. This argument is particularly poignant for this study as the problem that first drew me to Deleuze's philosophy is one that I identified during my professional experiences as a practicing architect. In the first chapter of Part 1, I will delve into one experience from my professional career to reveal a discrepancy between two ways of making assessments in design and regulatory practice: an assessment undertaken during and *within* the design and development process, and an assessment undertaken outside of the design and development process. As I will show later in Part 1, these two ways of thinking about assessments can be distinguished in several different ways, but the one that I find most compelling looks at the philosophical ideas used to create the two respective methods – namely, an essentialist approach and a non-essentialist approach. As Part 1 develops, I will show how and why I believe Deleuze's philosophical concepts can contribute to a formal assessment method constructed around the latter, and, thus, meet some of the aspirations of pragmatic planning theorists such as Stein and Harper (Stein and Harper, 2012), whilst capturing the concerns of discourse analysts like Fainstein (Fainstein, 2000) and the broader ontological arguments put forward by realists such as Van Wezemael (Van Wezemael, 2012). In doing so, this first part of the book sets the scene for an exploratory study into Deleuzian philosophy and how it might contribute to the tools used in planning.

1 A problem with assessments

> *Lodge yourself on a stratum, experiment with the opportunities it offers, find an advantageous place on it, find potential movements of deterritorialization, possible lines of flight, experience them, produce flow conjunctions here and there, try out continuums of intensities segment by segment, have a small plot of new land at all times.*
>
> *(Deleuze and Guattari, 2004b: 178)*

Identifying a problem '. . . par le milieu'[1]

In 2007/8 I acted as Project Architect for a proposed apartment block situated on the English Channel. This was one of many schemes in which I noted a clear discrepancy between the way members of the design and development team formed assessments and decisions as part of the design process, and the way an appointed sustainability assessor judged those decisions. The following overview of the scheme illustrates this discrepancy and why I believe it to be problematic and worth pursuing as a line of enquiry.

The apartment block was to be designed for wealthy downsizers from London and was supported, in principle, by the Local Planning Authority, who saw the scheme as a valuable contribution to a broader regeneration strategy for the area. The apartments were designed as a contemporary and luxury development using a steel frame structure, with concrete beam and block floors, and a flat roof concealed at ground level by a parapet. The external façade was formed from rendered panels, and slit glazing afforded views to the sea. Given the contemporary style of the development, the design team undertook an extensive process of public consultation, which resulted in a range of design changes. One year after commencing the initial sketch proposal, the scheme was awarded planning approval under delegated powers. As with most schemes of this size and type, planning approval was subject to a list of conditions. One of these conditions noted that the scheme would need to meet Code for Sustainable Homes level 4 (see sub-section 'The Code for Sustainable Homes' below).

The following text details the design decisions made in two particular areas of the scheme: the foundation and the structural frame.

Designing the foundation

After achieving planning approval, the project team began the detailed design stages (Royal Institute of British Architects work stage 4).[2] I started this process by commissioning a survey to identify the structural stability of the site. The survey revealed deep layers of sandy ground below the proposed building footprint. This sandy ground was the result of several key factors: the relief of the land, the kind of stone that formed the land and the corrosive tendencies of waves from the English Channel. In terms of the design, sand offers poor structural stability for a building proposal designed to be over five storeys above ground level and one storey below ground level. This discovery 'lodge[d] [the design team] on a stratum' and set in motion a series of 'experiments'[3] intended to resolve the 'design problem'.

The structural engineer suggested that designing the building around a deep piled foundation best accommodated this constraint. This provided us with 'an advantageous position' from which to explore three design options: steel driven piles, pre-cast concrete driven piles and contiguous piling. Each option was explored as a design sequence by experimenting with different materials, methods of construction and the entities that formed them.

The first option, steel driven piles, was deemed unsuitable due to the risks of corrosion from a high water table and the building's proximity to the sea front. A second option was to use pre-cast concrete driven piles. Whilst concrete piles would not corrode, they introduced other design and construction issues. This option would demand that the contractor excavate a large area of land, drive the piles at regular intervals, build a concrete block wall between the piles and treat the wall with a water-resistant membrane before back filling around the perimeter. This sequence of work would have had an impact on the estimated construction programme, the number of specialist trades needed, the amount of working space needed around the foundations and so forth.

A third option was to use contiguous, in-situ poured concrete piling. In this option, piles would be driven around the perimeter of the building and then filled with in-situ concrete and mixed with a water-resistant admixture to avoid subsequent water proofing. Once the concrete was set, the ground within this perimeter could be safely removed to form the basement level. This third option would have less impact on the programme, introduced fewer specialist trades, avoided risks associated with corrosion and could be adapted to suit the building footprint and the ground conditions below. Taking these points into consideration, the design team agreed that this third option was the most efficient and effective design solution.

Designing the structural frame

This decision led us to consider the design of the structural frame: a second 'stratum' in which to 'lodge' ourselves. The sketch scheme had been conceived using a steel frame. However, as with the steel driven piles, this steel frame would be at risk of corrosion from airborne salt from the sea.

We agreed, therefore, that the most effective method for reducing the risks of corrosion was to treat the steel frame with a protective paint. From this 'advantageous position', we were able to experiment with a range of different design and construction management opportunities and restraints. After further consideration, it was noted that, if this treatment was undertaken on an erected frame, there was a significant potential that connections, both to the concrete foundation and between steel sections, would be only partly covered. Discussion with steel fabricators provided us with a number of other opportunities. Anti-corrosion treatment could be applied off site, and special measures undertaken to avoid chipping the paint during transit.

Erecting the frame and subsequent phases of construction were less easy to resolve. To avoid chipping the paint during construction, the design team worked with the principle contractor, the curtain wall specialists, the cold steel partition specialists, and the cladding panel and insulated render specialists to consider a number of changes to aspects of the frame and methods of erection. These included the use of pre-cast fixing angles, protective covers, heightened site supervision and so forth. Such measures only partly resolved the corrosive tendencies of airborne salt and affected other parts of the design, adding further complexity to the design and construction process.

The risk of damaging the protective paint was all the more pronounced when we considered the subsequent maintenance of the building. The design team explored different options for replacing cladding and windows panels. These revisions to the detailed design demanded a number of bespoke materials installed by a range of specialist trades over a much longer construction programme.

With this in mind, the design team agreed to explore more fundamental design changes – namely, the use of a concrete rather than a steel structural frame. Unlike steel, concrete has a high resistance to the corrosive tendencies of airborne salt, thus avoiding the problems associated with specialist paints. Members of the design team noted other advantages and limitations resulting from this design change.

If the concrete frame was cast in-situ, it could be designed as an extension of the concrete foundation, thus avoiding vulnerable connections. This solution also reduced the number of trades working on site at any given time and reduced the amount of the transport associated with its construction. A concrete frame has better thermal and acoustic properties than a steel frame, thus reducing the demand for thermal and acoustic insulation. However, concrete columns take up more space than an equivalent steel column. Thus, to maintain the position and overall design of the windows, the columns would need to be positioned further into the building. This had implications on the proposed furniture layouts. It also increased the cantilever needed for the external balconies, and, thus, the thickness of the concrete floor needed to achieve this cantilever. This thicker floor would result in a deeper floor zone and a lower floor-to-ceiling height. Whilst this was not problematic in terms of intended use, it increased the space between floors, which impacted on the stair design. Our design 'experiment' with a concrete frame instead of a steel frame led us to re-design the stair by introducing more steps, which demanded more space. This had a knock-on effect on the space planning of the apartment corridors, the position of the lift and the layout of post boxes at ground level.

These two connected extracts from the design process highlight an on-going sequence of assessment and experimentation, not dissimilar to the creative process described by Deleuze and Guattari in the quotation introducing this chapter. It shows how designers ‘lodge themselves on a strata’ by identifying a key area of the design (such as the foundations and the structural frame) and identifying factors that influenced this design (such as sandy ground conditions and airborne salt). It shows how they ‘find an advantageous position on it’ by setting out a design proposal (such as the intention to introduce deep pile foundations formed in steel or concrete) and use this to ‘experiment with the opportunities it offers’ by assessing how sandy ground and airborne salt affect these design options before exploring a sequence of design changes that responded to these factors (changing the material or position of the structure or introducing protective paint, welded angles, protective sheeting, additional stairs and so forth).

However, this was not the only form of assessment applicable to this scheme. At pre-conceived stages in this process, the design was also assessed according to formal criteria, including those set out in the Code for Sustainable Homes.

The Code for Sustainable Homes

In the UK, the Building Research Establishment (BRE) developed the Code for Sustainable Homes as a building-industry-approved measuring tool for the concept of a ‘sustainable home’ alongside other BRE methods known collectively as BREEAM. In 2007 it became a supplement to Planning Policy Statement 1, ‘Delivering sustainable development’ (DCLG, 2005), published by UK central government, and it was applied to most housing developments until it was ‘wound down’ in 2015. This assessment, known colloquially as ‘the Code’, identifies nine categories of sustainable design. Each of these categories is broken down into contributing design issues (Table 1.1).

These categories and issues are weighted to reflect their respective contribution to the concept of the ‘sustainable home’. By comparing these factors against the properties of a design, an appointed Code for Sustainable Homes assessor is able to assign points and, thus, classify the scheme as more or less ‘sustainable’.

In the project outlined above, the assessor’s report discussed the decision to change from steel piling to contiguous concrete piling and the decision to change from a steel frame to a concrete frame. Because concrete produces higher CO_2 emissions than a comparable unit of steel, the assessor suggested that these changes to the design reduced the building’s fabric energy efficiency. As the category, CO_2 emissions, is weighted to account for 34.6% of the total sustainability assessment, this decision was deemed highly ‘unsustainable’. The report suggested re-considering this design decision, or to re-design other areas of the scheme to better reflect the criteria.

These conclusions and recommendations in the assessor’s report highlight an interesting difference between the formal method of assessment developed

Table 1.1 The Code for Sustainable Homes categories and issues

Category	*Issue*
Energy and CO_2 emissions	• Dwelling emission rate (M) • Fabric energy efficiency (M) • Energy display devices • Drying space • Energy labelled white goods • External lighting • Low and zero carbon technologies • Cycle storage • Home office
Water	• Indoor water use (M) • External water use
Materials	• Environmental impact of materials (M) • Responsible sourcing of materials – basic building elements • Responsible sourcing of materials – finishing elements
Surface water run-off	• Management of surface water run-off from developments (M) • Flood risk
Waste	• Storage of non-recyclable waste and recyclable household waste (M) • Construction site waste management • Composting
Pollution	• Global warming potential (GWP) of insulants • Nitrogen oxide emissions
Health and well-being management	• Daylighting • Sound insulation • Private space • Lifetime Homes (M)
Management	• Home user guide • Considerate Constructors Scheme • Construction site impacts • Security
Ecology	• Ecological value of site • Ecological enhancement • Protection of ecological features • Change in ecological value of site • Building footprint

(DCLG, 2010: 10)

by the BRE and the assessments made as part of the design and development process. By drawing on a universal set of criteria for a 'sustainable home', the assessor was unable to appreciate *why* steel was rejected in favour of concrete. In other words, this formal assessment method did not take full account of the integral and on-going approach to assessment used within the design process as noted above.

Second, it is questionable whether re-visiting the design using a steel foundation and structural frame as per the assessor's advice would have resulted in a more sustainable building. Whilst concrete does have higher CO_2 emissions than steel, the extract above revealed a number of other implications that might be deemed less 'sustainable' when considered against a broader definition of the term.

For example, a steel solution would have introduced higher levels of maintenance, and, thus, higher fabric-related CO_2 emissions over the lifetime of the building. It would have also demanded far more time on site, longer road closures and more bespoke materials each made and installed by different specialists coming from a much wider geographical area. The building would be at far greater risk of corrosion, which may have affected its durability, and constant maintenance could have affected the amount of time residents were prepared to stay in the apartments.

Issues with the BRE's 'universal method'

In 2011 the UK government set out its proposals for a 'Red tape Challenge' in response to growing concerns across the construction industry that assessments like the Code for Sustainable Homes and the regulations in which these assessments operate are not in keeping with the demands of design and construction practice (UK Government, 2011; DCLG, 2012a). In 2014 the decision was made to 'wind down' the Code for Sustainable Homes form of assessment. However, one should not assume that the issues I experienced in 2007/8 have been resolved. I have since acted as Project Architect for a number of other schemes in other sectors and have found the same disparity between BREEAM assessments and the assessments made during the design process. As in the example noted above, this disparity arises because these formal assessment methods are unable to take account of the reasons why decisions are made during the design process.

This continuing problem within specific milieu is not surprising. The Code for Sustainable Homes was developed by the BRE as an adaption of their previous assessment method, 'EcoHomes'. The BRE developed the CSH using the same universal method as all other BREEAM forms of assessment, including BREEAM New Construction. As these assessments apply to a great deal of projects undertaken in the UK across all principle sectors, we should not treat this problem as something limited to professionals operating in housing.

This is visible in Schweber's study of BREEAM in which she interviews a number of design professionals working in different sectors. Schweber's (2013) study, for example, is formed from a selection of interviews with key design professionals working on eight schemes formally assessed using the BREEAM assessments. These interviews highlight instances when design measures were introduced to meet universal sustainability criteria. These included the addition of

> refrigerant leak detection, flood risk surveys and flood alleviation measures, reduction in available parking spaces, cyclist facilities, and additional drinking fountains.
>
> (Schweber, 2013: 137)

Many of the practitioners interviewed by Schweber suggested that these changes resulted in questionable 'trade-offs' within the design as a whole (Schweber, 2013: 137). The respondents in Schweber's study traced these 'trade-offs' to a discrepancy between an official, universal understanding of 'sustainability' and the many different interpretations of 'genuine sustainability' (Schweber, 2013: 137). Many respondents explained these 'genuine' forms of sustainability as part of the design process.

The concerns expressed by construction professionals within Schweber's study are captured in a number of other studies. Cole's study, for example, criticises BREEAM for enforcing standardised design solutions over innovative responses to a given context (Cole, 2005). Others have argued that 'trade-offs' similar to those identified in Schweber's study can result in a poor representation of the building design as a whole, because higher scores in some areas of the design can be offset against lower scores in other areas (Sawyer et al., 2008).

Equally, we should not think of this as a UK problem as BREEAM is not limited to the UK. BRE Global reports an ever-expanding influence in other countries through the development of an international form of the BREEAM assessment.

Nor should we think that projects assessed using other methods are necessarily free from these issues. Whilst BREEAM is widely acknowledged as the leading assessment internationally, its principle rival, LEED, is designed around a very similar 'universal method'. LEED was developed in the USA but, like BREEAM, has been exported around the world and is now used in a number of schemes worldwide. Thus, it seems that the problems I identify from my experiences with BRE assessments affect many professionals working in different sectors and in different countries around the world.

The distinction between two kinds of assessment

So, the problem, it seems, lies in a distinction between two kinds of assessment: those made during/within the design and development process, and those made outside of the design and development process. From the perspective of design and development professionals, this distinction suggests that *certain* kinds of formal assessment – i.e. highly structured, universally applicable assessments like the Code for Sustainable Homes or BREEAM – (i) ignore many of the issues deemed important to actors outside the regulatory framework and (ii) result in decisions that such actors deem to be counterproductive to the values directing the assessment (such as achieving a sustainable design).

The question one might pose, therefore, is whether these concerns about formal assessments are limited to actors working within design and development roles (actors whose work is subject to assessment) or whether these concerns are equally felt by those working in regulatory/assessment roles (those responsible for forming and undertaking these assessments). To answer this question, I turn now to some of the debates within the field of planning.

It has long been argued that assessment is an intrinsic part of planning theory and practice. This stance is captured by Khakee, who notes that 'from both a theoretical and practical point of view "planning" and "evaluation" are inseparable concepts' (Khakee, 1998: 359). Whilst this assertion has gone mostly uncontested in planning literature, the nature, form and objectives of evaluation have been the subject of debate since the 1950s. The crux of this debate rests on the normative assumptions used to construct goal-centric, universally applicable assessment tools (similar to the Code for Sustainable Homes and BREEAM), and whether or not these assumptions continue to be relevant, or desirable, in contemporary evaluative practice (Borri, 1998).

Several attempts have been made to categorise different assessment methods in order to capture this debate. Khakee, for example, provides a historic review of planning assessments by drawing on and adapting Guba and Lincoln's (1989) classification of assessment methods (Khakee, 2003; Alexander, 2006). This classification identifies four 'generations' of assessment design. The first three generations are presented as a sequence in which the focus of assessment changes from a unit of measurement (such as cost), to pre-defined goals against which variations are judged (such as the goal for social equality or sustainable development in the case of BREEAM), to a greater focus on objective 'value free' standards (Khakee, 2003; Alexander, 2006). Despite this shift in focus, Khakee notes that these three generations of assessment design are underpinned by the same normative assumption – namely, that a single assessment model can be applied in a systematic manner to every project or policy used in all contexts.

It is this idea of a universal model of assessment that is challenged in the most recent 'fourth generation' of assessment methods. Rather than an assessment based on a pre-defined focus or set of criteria, such assessments

> take the form of a discourse between stakeholders who are directly or indirectly affected by a policy measure. It becomes a question of negotiations rather than of a search for a specific objective.
>
> (Khakee, 2003: 343)

This description of a 'fourth generation' of assessments resonates with the kind of assessments undertaken during the design process, as described in the example I used to open this chapter.

We can see a similar distinction between universal assessments and what I term 'assessments within the design process' or what Khaki terms 'fourth generation' assessments in the categories outlined by Söderbaum (1998). Söderbaum identifies three groups of assessment: highly aggregated assessments, intermediary assessments and highly disaggregated assessments. According to Söderbaum, the first of these assessment types, highly aggregated assessments, are characterised by their attempts to sum up the impacts or success of a project/policy according to a single value. As Söderbaum notes, such assessments assume that a specific set of values is common across all actors and areas of society (Söderbaum, 1998: 53). Like highly aggregated assessments, intermediary forms of assessment focus on one objective but aim to include the interests of different actors in the given project or policy.

The third kind of assessment, highly disaggregated methods, provide a clear shift away from the first two forms by rejecting the use of pre-defined objectives or assessment models intended to capture this objective:

> Since the aim is not to capture a single collective value or to reduce effects to a simplified scheme, these methods are multi-dimensional and flexible. The design of the methods adapts to the changing context, and not only are the results important but also the way of arriving at them.
>
> (Khakee, 2003: 345)

Whilst expressed in different ways, these two attempts to categorise the ideas underpinning assessment methods reflect a distinction between two broad groups of assessment design: an assessment that focuses on a single objective and uses a pre-defined, universally applicable method for assessing that design; and a flexible, contingent method that looks to identify the objective and approach of the assessment by engaging with the views of different actors in the project/planning process (Borri, 1998; Khakee, 2003; Alexander, 2006). According to Borri, developments in planning assessment theory over the last few years have been marked by growing support for the latter. This is particularly evident in academic literature, which shows growing concern about formal evaluation exercises (Borri, 1998). Such literature has questioned the normative assumptions held by such assessments, calling instead for a more 'sensitive' approach to evaluation 'in which the positions of the less politically represented – both humans and non-humans can be fully acknowledged and respected'. Such arguments have also suggested that combinations of expertise and 'commonsense' have resulted in more creative outcomes and greater levels of resilience (Borri, 1998: 276). This shift towards flexible forms of assessment that better reflect the views of different actors is often characterised as part of a broader paradigm shift across planning theory, noted by many as a move from positivism to post-positivism (Henkel, 1991; Khakee et al., 2000; Khakee, 2003) or/and from modernism to post-modernism (Alexander, 1998; Lawrence, 2000; Fischer, 2003).

This argument, and the debate within the field of planning assessment design, reflects the two observations I made above – namely, that there are two kinds of assessment: formal, universally applicable assessments like the Code for Sustainable Homes and BREEAM, and the kind of 'flexible' assessments made by different actors during the design (and development) process. As in my experiences, these arguments from planning literature suggest that the former is problematic because it (i) ignores many of the issues deemed important to actors outside the regulatory framework and (ii) results in decisions that some of these actors deem as counterproductive to the values directing the assessment. Rather than removing all formal assessments, therefore, the debates in planning theory suggest that these concerns might be resolved through the introduction of new, more context-specific assessment tools.

This shows that there is a case for developing a new, flexible form of assessment. Yet, whilst this case continues to grow in academic circles, it remains underdeveloped in practice. Alexander, for example, notes:

> In planning practice, if not so much in prescriptive theory, all the evaluation methods that have ever "evolved" are still in use today and those we think the most "primitive" are the ones enjoying the most widespread application.
> (Alexander, 2006: 14)

The evaluation theorist and practitioner Mary Henkel supports this observation. She notes that, despite theoretical shifts away from positivism, evaluative practice remains heavily reliant on positivist principles (Henkel, 1991).

Many attempts have been made within planning theory to explain why these flexible forms of assessment remain underdeveloped in practice. For Borri, it is owing to the influence of market-driven processes (Borri, 1998). For Lichfield and Prat, it is driven by governmental procedures that demand clear, demonstrable progress against policy objectives (Lichfield and Prat, 1998). Whilst for Baum, it is encouraged by the planners themselves, who demand 'psychological reassurances' of 'good' decision making (Baum, 1996; see also Alexander and Faludi, 1989).

I believe that the most compelling explanation for the relative absence of flexible, context-specific assessment tools in practice is captured in Fischer's support for a structured, universally applicable, systematic process underpinning the formal Environmental Assessments used at the scale of strategic planning (Fischer, 2003). To appreciate Fischer's arguments, and to help articulate the debate between advocates of both forms of assessment design, the following sub-section sets out the key points in Fischer's argument and follows this with a response from Richardson (Richardson, 2005).

Arguments for and against flexible, context-specific assessments

We can summarise Fischer's stance against flexible assessments into four arguments. The first of these arguments centres on the existing structures and policy advances currently in place. The objectives used to direct a systematic assessment process, he argues, are based on established definitions of the concept 'sustainable development'. This concept has been debated and discussed in policy literature for many years, and the resulting definition represents a consensus between these different political interests. The call for wide-ranging, context- and actor-specific definitions, he argues, works counter to these efforts and achievements. With this in mind, Fischer suggests that a flexible, context-specific approach to assessment would not be able to serve what is formally recognised as 'Sustainable Development'. Fischer's point concerning universal definition and consensus is not limited to this one concept. It could be equally applied to other broad, aspirational concepts (such as achieving social or territorial cohesion, a sense of community and so forth).

Fischer's second argument extends the idea of a single consensual definition by relating it to another universal definition: the idea that there is an underlying 'common-ground' in and across different parts of the development hierarchy. Reiterating the post-modern image of a society formed from conflicting systems of value, Fischer questions whether any method is truly capable of capturing such complexity and uncertainty. Rather than form a partial representation of this complexity, he argues, it is better to introduce a single set of values (a common good) that can help to unite different points of view. For Fischer, a systematic, object-led and universal method, like the one used in Environmental Assessments (EA),

> helps [to] reconcile differing goals and objectives through integration, thus uncovering inconsistencies and providing a platform for suggestions on how to achieve [a singularly defined objective for] sustainable development.
>
> (Fischer, 2003: 164)

This unifying approach, he adds, introduces benefits that would otherwise be absent from a more flexible and context-sensitive method. One such benefit includes the level of accountability offered by this approach to assessment design. By using the same method and principles at each stage in the policy, planning, and development process, one can map progress and identify blockages, deviations and improvements to this overarching agenda (Fischer, 2003).

Fischer's third argument rests on the idea that some degree of universality and systematic design are essential to all assessments. Even the most flexible post-modern methods, he argues, still rely on 'similar procedural stages as those present in "traditional" impact assessments' (Fischer, 2003: 163).

The fourth line of argument in Fischer's paper develops around his assertion that actors working in specific contexts are unable to pay due consideration to broader objectives. Drawing reference to the commonly used term 'NIMBY' (Not In My Back Yard), he suggest that many actors with a vested interest in a specific locality will place values specific to that locality, or their role in that locality, above those that he regards as national or international in scope such as agendas for sustainable development. In this sense, Fischer suggests that systematic, objective-led and universally applicable assessments like the Environmental Assessments are far more effective at capturing the broader 'public interest' than flexible, context-specific assessments.

Richardson's response to Fischer's paper is framed around these final two arguments. Whilst Fischer suggests that universal, systematic principles are an inevitable part of assessment design, Richardson argues that flexibility and differing values affect how such assessments are implemented. Building on a number of experiences working with Environmental Assessments in practice, he concludes that

> most actors – planners, politicians, and stakeholders – see EA as an opportunity to persuade, to mediate, and to contest. This is necessary because that is how planning is. EA is being used by actors in the real world to mediate and

> contest value differences and conflicts, and EA practitioners are engaged in the daily business of mediating values, consciously and ethically or not, yet the EA community seems divided on whether this is happening, appropriate, or desirable.
>
> (Richardson, 2005: 359)

This quotation shows that, for Richardson, efforts to construct scientifically sound procedures using fixed processes and definitions are not carried forward in practice. Whilst used to draw very different conclusions, this observation is equally acknowledged by Fischer, who, in previous studies, talks about a slippage in the use and understanding of EAs since their introduction in the 1980s (Fischer and Seaton, 2002).

For Richardson, this practical reality of systematic, objective-led assessments reflects a much broader issue in the principles used to construct them. In response to Fischer's fourth argument, Richardson challenges the underlying belief that 'bottom up expressions of (different) positions (will) interfere with broader, more strategic environmental aims' (Richardson, 2005: 347). Richardson's challenge rests on several connected points.

First, he argues that Fischer's comment suggests that local values act as the greatest barrier to creating a 'good' environment. Yet, he notes, one might equally point to a number of instances in which much broader, economic strategies have had a significant and negative effect on environmental conditions in the area. Second, it implies that one system of values (held by policy makers or/and expert planners) is more important than all other contradictory systems of value (held by many different actors with different roles and interests in a given context). The problem here, he argues, is that

> we do not yet have an accepted basis in planning for asserting or deciding that certain environmental objectives should, in certain cases, override locally expressed objectives.
>
> (Richardson, 2005: 348)

For Richardson, references to a political consensus on the definition of 'sustainable development' do not resolve this issue, because they still fail to explain *why* this goal should be prioritised over other competing goals. Third, Richardson asks whether it is ethical to 'simply design these supposed opposing positions out of the process' in order to create a method that better reflects a policy and development hierarchy rather than 'deal(ing) with the presence of multiple and often conflicting values, and ways of valuing' (Richardson, 2005: 348). For Richardson, an inability to capture these different views should be seen as a failure in the assessment design rather than as proof that such views should not be accommodated.

This academic debate between Fischer and Richardson reveals important lessons about the two kinds of assessments identified earlier in the chapter – namely, a formal, universal, goal-centric assessment and a flexible, multi-actor assessment undertaken during the design, development or planning process.

Whilst Richardson provides a compelling argument against two of the key points in Fischer's support for the former, he does not dismiss Fischer's underlying point that structured, systematic forms of universal assessment perform functions that are not easily met by a more flexible, contextually sensitive method. These kinds of assessment, like the EA, provide actors higher up in the development hierarchy with an indication as to how a given scheme relates to broader policy objectives, however questionable these might be. As Fischer notes, this structure and policy framework is already in place and is unlikely to change in any significant way over the short to medium term. Fischer's paper also demonstrates that these kinds of assessments provide a means through which different projects and interventions can be understood as part of a 'whole', rather than as a collection of individual cases. This brief review of the debate suggests that both kinds of assessments perform different roles and respond to different concerns.

Given that these assessments perform different roles, it is questionable whether one set of assessments could completely replace another. In other words, developing a flexible, context-sensitive assessment to fulfil the gap in practice noted above remains an important line of development, but one should not necessary expect that such assessments will fully replace existing, objective-led, universally applicable forms of assessment. This review of the debate also shows why proposals to integrate the two assessment types might prove problematic (see Lichfield, 2001 for example). As Richardson notes (2005), arguments for the former were introduced to challenge existing assessment practice and are based on very different principles.

Perhaps the strongest direction for developing assessment methods in light of these two stances is made by Alexander, who encourages efforts to develop new flexible methods, not to replace or merge with objective-led, universally applicable assessments but to offer planners a broader set of assessment tools (Alexander, 2006). This, he adds, will facilitate assessment selection to best reflect the demands of a given project and the demands of different actors rather than reflecting normative practice:

> it will be a rare case in planning or policy making when good practice will not call for a combination of some methods, in view of the limits of each particular evaluation method . . . understanding the whole kit of evaluation tools, and knowing which ones to select for a particular purpose, may be more important today than deeper familiarity with one family of methods or more sophisticated quantitative or statistical skills.
>
> (Alexander, 2006: 50)

Alexander's stance is a reasonable one given the points and observations discussed in this chapter. This leaves us with a chain of questions left unanswered, such as the following: are the kind of flexible, context-sensitive assessments outlined above possible? If so, what might they look like, and is there any validity in the suggestion that they should be developed alongside universal assessments?

These fundamental questions will be considered across the course of the following chapters and discussed specifically in Chapter 9.

So how do we go about making such assessments? A number of key figures in planning literature suggest that advances in planning theory might provide the strongest starting point. Lawrence, for example, suggests that the development of objective-led, universally applicable assessment methods, such as the Environmental Assessments, the Code for Sustainable Homes and BREEAM discussed above, have

> largely failed to benefit from planning theory insights and lessons. Obstacles and dilemmas already encountered and addressed in planning theory are still hampering [assessment] theory building and practice.
>
> (Lawrence, 2000: 607)

Whilst Richardson supports Lawrence's argument for a stronger link between developments in planning theory and assessment practice, he questions Lawrence's proposals, which he likens to ' "downloading" (complex theories) into a related area of practice, such as environmental assessment'. Instead, Richardson argues that one must, 'engage in more detailed exploration of the issues that are raised by particular theoretical approaches, so that the potential contributions can be handled sensitively' (Richardson, 2005: 342).

A similar argument is raised in much broader debates concerning the link between planning theory and planning practice. One of the key voices in this area comes from the planning theorist Heather Campbell, who notes:

> Few, if any . . . philosophers had planning in mind as the context in which they intended their ideas to be applied. . . . Their horizons are therefore broader than a subsection of public policy making concerned with the creation of place and the mediation of space. This is not to dismiss this hugely significant body of work but to caution the appropriateness of direct translation.
>
> (Campbell, 2006: 93)

Campbell and Richardson, therefore, share in common the idea that abstract theory, especially theory with a strong link to philosophy, cannot be transferred directly into planning practice. 'Rather than seeking to import "lessons" from planning theory', argues Richardson, 'the aim is to try and "see" [assessment practice] through the eyes of planning theory' (Richardson, 2005: 342). In doing so, one must use theory in a different way:

> instead of advocating solutions to procedural dilemmas, for example, the idea is to see if these dilemmas might be thought about differently if they were addressed in the ways that some planning theorists have thought about planning.
>
> (Richardson, 2005: 342)

These arguments suggest that, to develop an alternative form of assessment that reflects some of the observations and arguments made above, one must draw on philosophy and planning theory as a way to 'see' problems, to critique existing normative practices, and to re-think how these practices might play out differently.

This line of enquiry provides useful insight in to the aims of this book – namely, to consider how Deleuze's philosophy might be of use to planning practice. Before introducing Deleuze into this problem, I begin Chapter 2 by uncovering and exploring the essentialist principles underpinning existing forms of objective-centric, universal assessments like the Code for Sustainable Homes and BREEAM. Building on these arguments, I will show why I believe Deleuze's non-essentialist concepts offer new opportunities for 'seeing' these concerns and developing a new kind of assessment tool that better reflects complex issues specific to context and the way professionals engage in that context.

Notes

1 I translate this as, 'Identifying a problem . . . from within the specifics of a field/situation'.
2 The RIBA work stages 1–8 are set out in the following link: http://www.architecture.com/TheRIBA/AboutUs/Professionalsupport/RIBAOutlinePlanofWork2013.aspx#.U2n1dfldV8q.
3 As per quotation used to introduce the chapter.

2 Essentialist and non-essentialist assessments

An essentialist phenomenon in planning assessments

As noted in the closing paragraph of the last chapter, the first aim of this chapter is to appreciate formal assessments like the Code for Sustainable Homes and BREEAM in broader, theoretical terms. To assist with this exercise, I start by re-presenting the table offered in the technical literature for the Code for Sustainable Homes in a spatial rather than a tabular format (Figure 2.1).

The figure below shows how the concept of the 'sustainable home' branches out into nine other concepts, and each of these concepts, in turn, branch out into a number of different factors. If we think back to the last chapter, 'the sustainable home' acts as a single goal broken down into a number of factors deemed fundamental to achieving this objective.

A very similar diagram can be seen in other assessments constructed by the Building Research Establishment, known collectively as the Building Research Establishment Environment Assessment Method or BREEAM for short (Figure 2.2).

Figure 2.2 shows just how closely the BREEAM method is to other BRE methods. It too defines a vague concept, in this case the 'sustainable, newly constructed building', into other concepts that are, in turn, defined according to specific factors. Indeed, if I were to set out a diagram for all the other methods in the BREEAM suite and remove the concept at the head of the diagram, even the most highly trained BREEAM assessor would struggle to tell the difference. This highlights an important feature of the BREEAM suite. It shows that concepts like 'the sustainable home' act as a single goal broken down into a number of factors deemed essential to achieving this objective. It also shows that these factors are essential not only to one specific kind of sustainable development but to sustainable development more broadly.

Using the assessment frameworks outlined in the last chapter, one might conclude that these methods belong to, what Khakee's termed, the 'second generation' of assessment methods because it is constructed around pre-defined goals against which variations are judged (Khakee, 2003). These figures also show that, by summing up the impacts or success of a project according to a single value (i.e. a Code

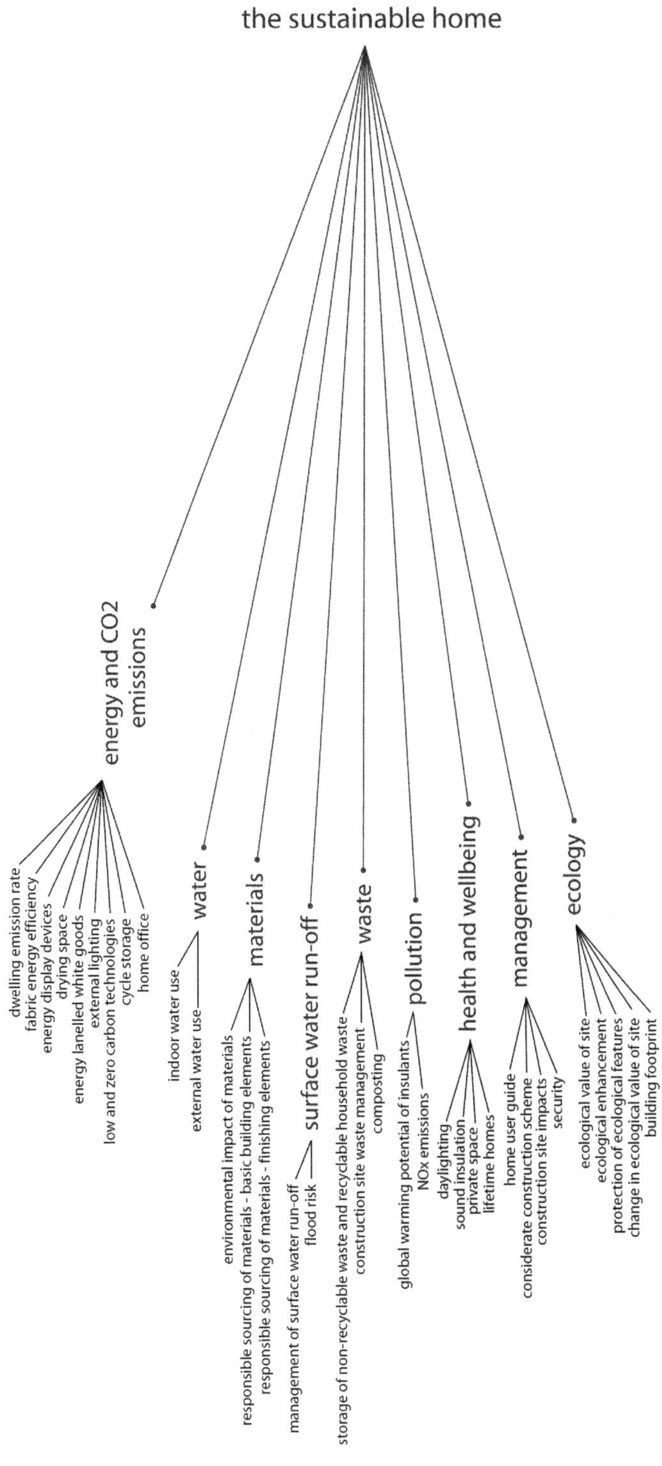

Figure 2.1 Diagram of the Code for Sustainable Homes

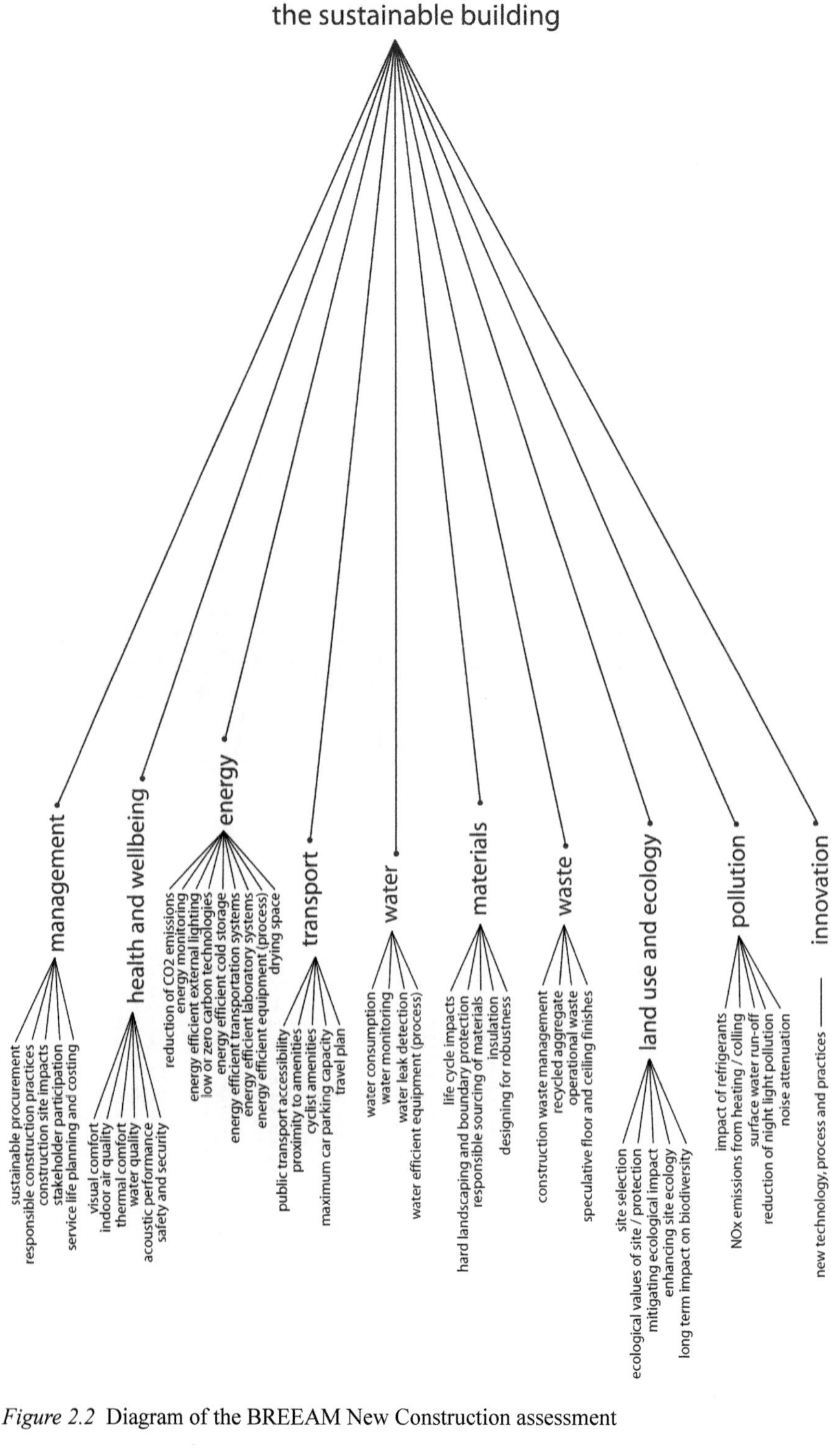

Figure 2.2 Diagram of the BREEAM New Construction assessment

for Sustainable Homes or BREEAM rating), this assessment reflects Söderbaum's description of a 'highly aggregated' form of assessment (Söderbaum, 1998).

These two frameworks help position the Code for Sustainable Homes and BREEAM amongst other assessment types within a select area of planning and loosely explain the principles used in their construction. However, as Richardson notes, if one is to 'see' the problems underpinning these assessments, and to explore alternative directions for development, then one must draw on more abstract, philosophical ideas.

With this in mind, I posit that the tree-like structure of BRE forms of assessment are the result of a set of principles drawn from an area of thinking known as 'essentialism' (see Abrahams, 2014). In simple terms, essentialism holds that any material or non-material entity has a number of traits that are deemed essential to its identity or/and its function (Cartwright, 1968; Sayer, 1997; Barrett, 2001). In this instance, Figures 2.1 and 2.2 above suggest that there are nine components that, together, form the total identity and function of a sustainable home[1] and ten components that, together, form the total identity of a 'sustainable newly constructed building'.

Whilst these components and traits may all be considered as essential, the Code and BREEAM do not treat them as equal contributors to this identity and function. In both figures above, I have positioned each component or trait at heights relative to their weighting. So, for example, in Figure 2.1, 'energy and CO_2 emissions', 'water' and 'materials' are all essential to the concept of the sustainable home. Yet, 'CO_2 emissions' forms 36.4% of the concept, 'water' accounts for 9% and 'materials' account for 7.2%.

The figures also illustrate how the same essentialist approach is used to define the identity and function of these components. There are, in Figure 2.1 for example, nine components that, together, form the total identity and function of 'CO_2 emissions'. And again, these components are weighted differently to reflect the extent of their contribution. The result of this essentialist approach is a universally applicable assessment 'model' for the otherwise vague concept of 'the sustainable home', the 'sustainable, newly constructed building' and so forth, which, for the purposes of this study, I would like to capture using the shorthand term, the 'modelled concept'.

Other essentialist modelled concept-based assessment tools

Whilst Khakee's and Söderbaum's frameworks were used to categorise a select area of assessments, the theoretical image of the 'modelled concept' helps to reveal a much broader essentialist phenomenon affecting the construction of many kinds of planning and design tools.

At the scale of European spatial policy, one can identify the same essentialist principles in the efforts made by the European Spatial Policy Observation Network to measure the concept of 'territorial cohesion' (European Territorial Cohesion Indicators [ETCI]) and to assess how spatial policies across the EU contribute to the concept of territorial cohesion (Territorial Impact Assessments [TIA])

(Abrahams, 2014). In both instances, the concept of territorial cohesion is modelled according to its essential traits.

At the national scale, the concept of 'deprivation' has been modelled many times in the UK over the last few years, each attempt resulting in a variant of the assessment tool, the Index of Multiple Deprivation (IMD). This interest in deprivation has been mirrored with concern over national 'wellbeing' and 'happiness' (ONS, 2011b). The Office for National Statistics has undertaken a number of research projects to measure these concepts (ONS, 2011a; 2011b). As with the ETCI and IMD, these multiple indicators are based on efforts to define a vague concept through what are believed to be its essential traits.

At the level of urban development, Heritage Impact Assessments are often used to assess a proposal's impact on a site of historic importance. The criteria for judging such impacts are determined by recognised standards for evaluating buildings or sites of Outstanding Universal Value (UNESCO, 2013). This measure is formed from various criteria used to award UNESCO status for World Heritage sites (ICOMOS, 2010). Whilst less strict than other modelled concept-based tools, these criteria and the assessments used follow a model drawn from the vague concept of 'Outstanding Universal Value'.

At an even finer scale, modelled concepts are also used to form sector-specific assessments such as the Housing Quality Indicators 2007 and the London Housing Design Guides 2010 (The National Affordable Housing Agency, 2007; Design for London, 2010). Both of these guides define a number of traits deemed essential to a 'well-designed' or 'quality' housing scheme. And both of these guides use these traits to identify material factors that serve as a definitive checklist for designers, developers and planners.

These examples show that 'modelled concepts' have underpinned the construction of many different assessment tools used by a range of actors in development and regulatory processes. Whilst my use of the term 'modelled concept' is a new addition to the study of such assessments, as I will show in the following sub-section, I am not the first theorist to make the link between essentialism and objective-led, universally applicable assessment tools. Similarly, I am not the first theorist to show that these essentialist principles result in tools that fail to reflect complex contextual factors and the decisions made by actors working within those contexts. Some of the most compelling and detailed arguments against essentialist assessment tools are presented by planning theorists belonging to two traditions: pragmatism and discourse analysis.

Debates in planning theory: considering alternatives to essentialist tools

Pragmatism and communicative planning

My concerns over essentialism and the tools derived from essentialism – i.e. tools based on modelled concepts – are reflected in the work of several pragmatist planning theorists. According to Healey, pragmatists share a mutual rejection of simple essentialist definitions used as 'first principles' in 'designing practice' (Healey, 2009b). And, as in the example used to introduce the problem from my experiences in practice in Chapter 1, many of these positions draw on cases from specific milieu to highlight the problems associated with essentialist planning tools.

Jamal et al.'s study shows how many disagreements and divisions during the planning process can be traced to the use of essentialist labels like 'the sustainable scheme' (2002). Some of the empirical studies outlined by their study suggest that such problems arise when the characteristics of a particular issue or stakeholder group do not translate into a pre-conceived definition (2002).

When these definitions are used as the basis for planning tools, it seems they introduce other problems. In their study of 'sustainable planning' in rural communities, Markey et al. found that these tools were incompatible with rural contexts and could not be adapted (2010: 19). Successful projects undertaken in the study area, they note, were those in which simple solutions could be found to meet problems as and when they arose (2010: 16, 19).

For many, such findings are not limited to any one geographical area or setting. Stein and Harper draw on empirical studies to identify conclusions about the effects of essentialism in planning practice more broadly. They argue that essentialist planning tools impede creativity and innovation because they encourage designers and planners to apply rules rather than respond to specific issues (Stein and Harper, 2012). These problems occur because of a disjunction between a politically motivated concept and the pragmatic demands that arise during implementation (Batty, 2006: 38). Whilst the former seeks certainty and the reduction of risk, the latter must engage with the ' "wickedness" of complexity and uncertainty' (Batty, 2006: 33).

To resolve these problems, many pragmatists have argued that normative planning practice should re-think its engagement with complexity and uncertainty. One of the most important ways of doing this, they argue, is by keeping concepts like sustainability 'fluid', 'fuzzy' and adaptable (Jamal et al., 2002; Harper and Stein, 2006; de Roo and Porter, 2007; Healey, 2009b; Stein and Harper, 2012). Many studies have shown that this fluidity is often unavoidable. Drawing on several planning studies in several countries, Healey notes that concepts constructed in one milieu change when they 'travel' to other parts of the world and into other planning systems (Tait and Jensen, 2007; Healey and Upton, 2010; Healey, 2012).

In this pragmatic alternative, therefore, truth or meaning is held as something constantly reasserted through experiences *within* practice. For Healey, planners must constantly ask, 'so what does this mean for us here' and now? (2009b). In this sense, meaning is considered temporal and immanent. Or, as Richard Rorty argued, the concepts we use are 'temporary resting places constructed for specific utilitarian ends' (1982: xli) rather than fixed and universally applicable.

Stein and Harper argue that these temporal, immanent concepts should be created through dialogue. Their proposals for a 'dialogical planning' approach build on developments in communicative planning and the techniques used to construct meaning within the specific conditions of an issue and context (Healey, 1996; Innes and Booher, 1999; 1999a; Harper and Stein, 2006).

Discourse analysis

Essentialism is seen as equally problematic for discourse analysts such as Hajer, who notes that discourse analysis is based on

> an anti-essentialist ontology; it assumes the existence of multiple, socially constructed realities instead of a single reality, governed by immutable natural laws. . . . [It] takes a critical stance towards 'truth' and puts emphasis on communications through which knowledge is exchanged.
>
> (Hajer and Versteeg, 2005)

However, whilst discourse analysts and pragmatists share a common 'anti-essentialist' stance, they approach this argument in a very different way. Indeed, discourse analysts provide one of the most convincing and consistent challenges to pragmatist and communicative planning.

For some political economists and key discourse analysts, pragmatism and communicative planning does not take sufficient account of broader, political forces (Flyvbjerg, 1998; Fainstein, 2000; 2005; Huxley, 2000) or conflict (Flyvbjerg and Richardson, 2002). According to Fainstein, it is 'naïve' for us to imagine that the meanings we construct through dialogue do not draw on wider political discourses and ideologies (Fainstein, 2000).

For political economists like Fainstein, concepts are understood through a range of meanings rather than according to one essentialist definition. But, unlike pragmatic-collaborative planners, they hold that these meanings are not created from *within* practice alone. Rather, these constructions are conditioned by overarching discourses. Hajer's study of acid rain in British politics (1993), Richardson and Jensen's study of space in European politics (2001) and Waterhout's study of territorial cohesion in European spatial planning (2007) provide good examples of this argument. These studies each identify the dominant meanings of a concept and link them to a set of discourses each competing for a hegemonic position in the field.

As analytical studies, these examples do not provide planners with an alternative, non-essentialist approach to practice. However, they do help the planning community understand the consequences that arise when one essentialist definition dominates practice. Richardson and Jensen found that, in order to embed itself within planning practices, a discourse must marginalise other discourses and other meanings (2001: 715; 2003: 19). By doing so, policy is directed towards one group of actors and away from another (Richardson and Jensen, 2001).

The empirical studies undertaken by pragmatist/collaborative planners and discourse analysts highlight consequences arising from the use of modelled concepts not identified in my own experiences as set out in Chapter 1. These include exacerbated conflicts in planning discussions, 'implementation gaps', suppressed creativity and marginalised interest groups. Bringing these studies together, pragmatism, collaborative planning and discourse analysis suggest that concepts should be considered as 'fluid' and adaptable. Their meaning should be created to meet the operational demands of practice (as argued by pragmatists). However, not all creations are possible. The possibilities for creating concepts are conditioned by broader forces, including those directed by politics and discourse (as argued by discourse analysts).

The problem with pragmatism/collaborative planning and discourse analysis

Whilst these arguments provide a useful stance against essentialism, and show a number of commonalities with my own observations and concerns, they are limited by their human-centric understanding of the complex world around us. The pragmatic/collaborative planner focuses on planning discussions, whilst the discourse analyst focuses on the conditions set by political discourse. Van Wezemael argues that this non-realist focus on language 'reinforce[s] a perspective where almost everything is related to hermeneutic, linguistic or sociological constructions'. (Van Wezemael, 2012: 97). And, in doing so, these stances suggest that the complexity seen in design/plan-making[2] processes is 'a property of humans interpreting reality' rather than 'a property of reality' itself (Van Wezemael, 2012: 97).

For Van Wezemael, the human-centric, interpretative approaches underpinning pragmatism/collaborative planning and discourse analysis maintain 'the essentialist perception of matter as inert, passive and docile' (Van Wezemael, 2012: 98). Thus, whilst these areas of planning theory have been keen antagonists of essentialist definitions, for Van Wezemael, they form these anti-essentialist arguments whilst holding onto an essentialist understanding of the material world. If we are to avoid material as well as conceptual essences, he argues, we must see 'complexity [as] a property of reality' and the design/plan-making processes as a way of engaging within this complexity (Van Wezemael, 2012: 97).

Van Wezemael's broader critique of essentialism, and his support for complexity as a property of reality, points to a new direction for developing arguments against the essentialist phenomenon in planning. The direction that Van Wezemael and others have taken in order to better understand complexity and essentialism returns us to the philosophical terrain that directs this book – namely, the realist philosophy of Gilles Deleuze (see DeLanda, 2006; Hillier, 2007; Van Wezemael, 2008; 2010; 2012; Wood, 2009; Hillier and Van Wezemael, 2012; Trummer, 2012). As Dovey notes, Deleuze's philosophy allows us to

> overcome simplistic [essentialist] divisions between materiality and meaning, architecture and planning . . . it enables us to break with static, fixed, closed and essentialist notions of place.
>
> (Dovey, 2013: 138)

Deleuze's philosophy

The versatility and breadth of concepts used in Deleuze's realist ontology offers some explanation for its growing popularity amongst the planning community. Deleuze constructs his philosophy by drawing on a number of different fields of knowledge. Such fields include art, cinema, literature, architecture, geography, psychoanalysis, politics and history. Each of these disciplines presents Deleuze with a different 'milieu' in which to construct and re-channel his ideas (Deleuze and Guattari, 2004b). Deleuze uses concepts from these different 'milieux' to serve very similar, explanatory roles in his ontology. In one such example, DeLanda argues that the 'intensive processes' drawn from his study of complexity science is used in a similar way to 'territorialisation' from his study of geography and geology, 'coding' from his study of linguistics and institutions, 'desiring production' from his study of psychoanalysis and the 'plane of composition' in his study of art and philosophy (DeLanda, 2002: 202–223).

In doing so, Deleuze achieves two things. First, he offers theorists from different fields of knowledge a number of gateways into his work. Second, he shows that the

> conclusions [he makes] do not depend on his particular choice of resources, or the particular lines of argument he uses, but that they are *robust to changes* in theoretical assumptions and strategies. Clearly if the same conclusions can be reached from entirely different points of departure and following entirely different paths, the validity of these conclusions is thereby strengthened.
>
> (DeLanda, 2002: 4)

For DeLanda, therefore, Deleuze's philosophical project intends to demonstrate how the theoretical framework he creates can be identified in, and subsequently used by, all areas of knowledge. Massumi captures a similar point in his preface to the Chinese edition of *A Thousand Plateaus* when he states that Deleuze's work provides a 'conceptual toolkit' to help resolve problems in different fields (Massumi, 2010).

So, what is Deleuze's non-essentialist ontology?

Before I discuss the ontological concepts Deleuze uses to oppose essentialism, I would like to explain how I have structured my reading of Deleuze's concepts. After all, there have been a number of attempts to reconstruct Deleuze's ontological framework. Levi Bryant, for example, reconstructs Deleuze's work and combines it with ideas from other theorists including Harman and Bhaskar to support his argument for an object-orientated ontology, or 'onticology' (Bryant, 2011). His later work represents a shift from the object to the machine as the centre of his ontological framework – the result being what he terms a machinic ontology or 'onto-cartography' (Bryant, 2014). In a seemingly opposing stance, Brian Massumi reconstructs Deleuze's work and combines it with ideas from James and Whitehead to support his argument for an event-orientated ontology (Massumi, 2013). These object, machinic and event-focused studies are important interpretations. But their connection to this study is less obvious than the one offered by Manuel DeLanda, who frames his reconstruction around a rejection of essentialism (DeLanda, 2002) and our understanding of space (DeLanda, 2006). Given that these are fundamental issues in my search for an alternative to essentialist assessment methods, I intend to call on DeLanda's work in particular and to draw on other interpretations as broader frames of reference.[3]

According to DeLanda, Deleuze's philosophy provides us with three arguments against essentialism and five concepts serving as the basis for an alternative approach. These concepts are the assemblage, becomings, multiplicity, blocks of becoming, and the virtual diagrams (plane of immanence).[4] Whilst DeLanda only makes passing reference to machines in his 2002 text, his previous study, *A Thousand Years of Linear History* (1997), positions these machines as fundamental to the processes through which beings are actualised. On this basis, the 'machine' will serve as the sixth concept.

As I will show in the following, Deleuze's arguments against essentialism and the concepts he creates to avoid essentialism provide a strong basis from which to understand the problems associated with modelled concept-based tools outlined in Chapter 1, and a useful direction for developing an assessment method not constructed around a modelled concept.

Argument 1: a whole and its parts

The first of Deleuze's arguments against essentialism concerns the relationship between a whole and its component parts. According to DeLanda, essentialism is characterised by an underlying belief in the

> relations of interiority: [in which] the component parts are constituted by the very relations they have to other parts in the whole.
>
> (DeLanda, 2006: 9)

Because of this relationship, essentialists assume that we can understand the whole by the cumulative properties of the components. And, inversely, the properties of the whole can be used to explain the properties of the individual components (DeLanda, 2006).

These relations of interiority can be seen in the modelled concept for 'the sustainable home' and the 'sustainable newly constructed building'. Figures 2.1 and 2.2 suggest that the sustainable building as a whole can be explained, and indeed is assessed, by the list of factors at the bottom of the respective figure. Equally, it suggests that the meaning of CO_2 emissions is explained by its position within the assessment model as a whole.

The Assemblage

Deleuze argues that these relations between the whole and their components fail to account for instances when the whole has properties not present in the component parts. A building may contribute to the sustainability of local businesses, for example, yet this property of the building as a whole might not be visible in the components that form the design. Equally, there are instances when individual components have properties that do not contribute to the building as a whole but contribute to other buildings, other spaces or other social groups. A building façade might be designed to include false windows to reflect local vernacular or to improve the safety of a community, even though these windows contribute little to the building's intended use. Thus, for Deleuze, not all properties are utilised by the totality in which we find them. And taking this further, entities should be understood not according to their properties but by the way they operate in combination with other entities in different groups (DeLanda, 2006).

This argument for the relations of exteriority is captured in Deleuze's suggestion that all entities are formed as a collection of other entities, which he terms '*the assemblage*'. Building on Deleuze's stance, DeLanda argues that each and every assemblage can be understood both as an individual formed of component individuals *and* as a component in another assemblage.

Argument 2: the referential model

This leads me onto the second argument Deleuze raises against essentialist thinking. According to DeLanda, this argument relates to the use of 'essences [that] . . . act as *models*, eternally maintaining their identity' and used to understand the individuals we come across in the future (2002: 6). For Deleuze, this approach starts from the wrong position. Rather than studying the real world to justify a set of definitions defined in advanced, he argues that we should start from *within* the real and the differences we find there (DeLanda, 2002).

This argument against the referential model is at the heart of Deleuze's seminal monograph, *Difference and Repetition* (2004). For Deleuze, the concepts of difference and repetition are some of the most underdeveloped in philosophy. In both instances, he argues, they are only understood in reference *to* a stable, identified case or model and judged according to the extent of their similarity. Difference, for example, is only ever understood as difference *to,* and seen as a negation of something already agreed upon (Deleuze, 2004).

The concept of repetition suffers from a similar problem. Repetition is always considered as the repetition *of* an original, identified case. This, he argues, is 'a static repetition . . . [one that] refers back to a single concept, which leaves only an external difference between the ordinary instances of a figure' (Deleuze, 2004: 23). For Deleuze, therefore, these repetitions recognise objects according to their similarities rather than their 'internal differences, which it incorporates in each of its moments' (2004: 23).

This critique of repetition and difference can be seen in the example I drew from my experience in practice in Chapter 1. For the Code for Sustainable Homes assessor, the decision to change from a steel frame to a concrete frame took the design further away from the pre-conceived referential model of a 'sustainable home'. In other words, this difference was judged as something different *from* the referential model of the 'sustainable home', not a difference generated from within the design itself.

Deleuze's concept of the assemblage helps us overcome this focus on sameness and similarity. For Deleuze, each assemblage we see around us in the actual world is unique, which is why he asks that we think of them as individuals (DeLanda, 2006). When we bring entities together, we are creating something new that responds to the different demands of each entity. The assemblage, therefore, avoids the risks of 'applying [universal] rules' identified by Stein and Harper (2012) and the risks of marginalising particular sets of entities such as interest groups discussed by Richardson and Jensen (2001; 2003).

Argument 3: properties

Deleuze's third argument against essentialism concerns *what* we should study. Essentialism assumes that we should study individuals as if they were fully formed entities defined by their properties. However, in doing so, Deleuze argues that we ignore the ways in which these individuals (or assemblages) come into being (DeLanda, 2002: 10). As a consequence, essentialism is unable to consider how assemblages might behave in the future and how they might change when confronted with other individuals in a changing environment.

For Deleuze, understanding how an assemblage comes into being demands a shift in the way we think about them ontologically. Rather than asking what an assemblage *is* (its being), Deleuze argues that we should think about the potentials that determine what it might become (its *becomings*). DeLanda captures this idea when he notes that

> becoming without being . . . [is one in which] . . . individual beings do not exist but only as the outcome of becomings, that is, of an irreversible processes of individuation [or actualisation].
>
> (DeLanda, 2002: 106)

This quotation suggests that, for Deleuze, reality is much more than the actual world we see around us. These 'pre-actual' parts of reality are fundamental to Deleuze's philosophy and contain some of his most complex ideas. Before I discuss becomings in further detail, therefore, I would like to provide a brief outline of this ontological structure drawing on the explanation used by DeLanda (2002).

Deleuze's ontological structure

One of the best ways to understand this ontological structure is to envisage a line between two extremes: an actual realm and a virtual realm (Figure 2.3). As DeLanda notes above, the first of these extremes, the actual realm, is problematic because it suggests that the entities we see around us can be studied as roughly fixed beings with discernible traits. A similar argument could be made against Deleuze's reference to an opposing, virtual realm. A collection of 'pure potentials' is, after all, no less abstract and reductionist.

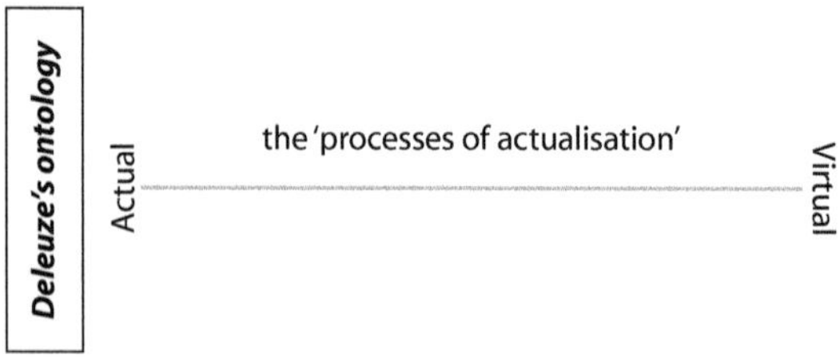

Figure 2.3 Schematic of Deleuze's ontological structure

Rather than afford these extremes with an ontological status, as is the case in essentialism, Deleuze sees these two realms as functional devices. They help him envisage the theoretical line connecting these two points, or what DeLanda has termed the 'processes of actualisation' (Figure 2.3).

This conceptual technique is not unique to Deleuze's ontological framework. Similar examples can be found in many of the concepts he uses to develop this framework. Notable examples include the concepts of the tree and the rhizome, territorialisation and deterritorialisation, the state machine and the war machine, striated space and smooth space and so forth. In each instance, these extremes are not intended as ontological truths. There is, for example, no such thing as an entirely rhizomatic or tree structure; neither is there such thing as a true state or war machine.

In each instance, these conceptual devices serve a similar role. They provide Deleuze with a middle ground in which to explore processes of actualisation. And, more broadly, they emphasise what DeLanda termed 'a common conclusion' in Deleuze's work (DeLanda, 2002: 4): that all assemblages formed from human and non-human, material and non-material entities should be understood according to a combination of actualised and dormant becomings, rather than as beings in their own right.

Becomings

So what are these becomings and what do they do? According to DeLanda, Deleuze's concept of becoming is drawn from advances in mathematics and physics. In such fields, becomings are referred to as 'singularities', which

> are said to represent the inherent or intrinsic *long-term tendencies* of a system, the states which the system spontaneously tend to adopt in the long run as long as it is not constrained by other forces.
>
> (DeLanda, 2002: 14)

Building on these scientific findings, DeLanda shows how the same singularity might not necessarily result in the same form (2002: 15; 2006: 29). DeLanda illustrates this by referring to the formation of soap bubbles and salt crystals. In this simple example, he notes that these two forms are the result of a common 'long-term tendency' towards the minimisation of energy use. The soap bubble adopts a spherical form to minimise surface tension, and a salt crystal adopts a cubic form to minimise bonding energy (DeLanda, 2002: 15).

In such examples, we can see why Deleuze contests the essentialist focus on properties instead of becomings. As DeLanda notes, an essentialist might look at the soap bubble and the salt crystal and conclude that these two entities belong to two completely different essentialist groupings. In doing so, this basic assessment would fail to acknowledge the same becoming that allowed these two forms to come into being (DeLanda, 2002: 15). This emphasises the problem raised in Chapter 1 – namely, that essentialist, formal assessments like the Code for Sustainable Homes and BREEAM are unable to take account of the design process and the assessments made during this process because they focus on the building as a complete being.

Multiplicities

There are, of course, significant differences between the formation of a soap bubble and the development of a building or urban design. However, for Deleuze, these types of differences are scalar rather than ontological. Unlike the soap bubble, a building assemblage is formed from many singularities, each pushing the design process towards different forms. Deleuze refers to this combination as the assemblage's 'virtual diagram' or, in scientific terms, as an assemblage's 'multiplicity' (DeLanda, 2002). We can see this multiplicity in the example outlined in Chapter 1. The two extracts discuss the design of the building's foundations and the building's structural frame. In each extract, I outline a number of different becomings, or long-term tendencies all contributing to the building assemblage: a long-term tendency towards corrosion, low maintenance, durability, reduced energy consumption, ease of construction, cost efficiency and so forth. Taken together, these long-term tendencies are part of the building's multiplicity, or 'virtual diagram'.

Blocks of becoming and force

So how might these becomings forming the multiplicity direct the design process? According to DeLanda, singularities (becomings) affect the process of actualisation at different stages (2002: 26). In each stage, development is directed towards a dominant or universal singularity, until it meets a threshold, which moves the process into a new block directed towards a different universal singularity.

In *A Thousand Plateaus* Deleuze and Guattari refer to these stages as 'blocks of becoming' (Deleuze and Guattari, 2004b: 324). Whilst it is possible to identify the universal singularity directing these stages, Deleuze and Guattari argue, it is not possible to know how this will play out, or at what point one block will shift to another. This is because 'there is no preformed logical order to becomings and multiplicities' but 'alogical consistencies or compatibilities' (Deleuze and

Guattari, 2004b: 276/7). Contextual factors may introduce new thresholds, or may steer a development towards an otherwise dormant singularity. In such instances, a new block of becoming begins, which may lead the developing assemblage towards a different form (DeLanda, 2002).

The concepts of becomings, thresholds and blocks of becomings appear abstract when discussed in such philosophical and scientific terms. To illustrate their usefulness in the field of design/planning, Figure 2.4 provides a marked-up extract from Chapter 1.

The figure below illustrates how one might use Deleuze's concepts to explain the sequence of design experiments and assessments made during the design process discussed in Chapter 1. The body of the text is formed from extracts taken from one of the design sequences: 'designing the structural frame'. The notes in the right-hand-side column translate aspects of this design process into Deleuzian terms.

This extract shows how design sequences were directed by blocks of becoming, which changed once they met a threshold resulting in a new direction (a new block of becoming). This extract also shows how the development of one block of becoming can introduce otherwise dormant becomings, such as a 'becoming thermally efficient of the structural frame' or a 'becoming simple of the construction process'.

Designing the structural frame. This decision led us to consider the design of the structural frame. The sketch scheme had been conceived on the basis of a steel frame. However, as with the steel driven piles, this steel frame would be at risk of corrosion from air-borne salt from the sea.	*BLOCK OF BECOMING: Becoming structurally stable* *THRESHOLD IN THE DESIGN: Becoming corrosive of air-borne salt*
The design team agree that the most effective method for reducing the risks of corrosion was to treat the steel frame with a protective paint.…Such measures only partly resolved the corrosive tendencies of air-borne salt and affected other parts of the design adding further complexity to the design and construction process...	*BLOCK OF BECOMING: Becoming corrosion resistant of steel*
The design team explored different options for replacing cladding and windows panels. These revisions to the detailed design demanded a number of bespoke materials installed by a range of specialist trades over a much longer construction programme.	*THRESHOLD IN THE DESIGN: Becomings high maintenance of the building, becoming complex of the construction process/ programme ...*
With this in mind, the design team agreed to explore more fundamental design changes, namely, the use of a concrete rather than a steel structural frame.…	*BLOCK OF BECOMING: Becoming concrete of the frame*
If the concrete frame was cast in situ they could be designed as an extension of the concrete foundation, thus avoiding vulnerable connections. This solution also reduced the number of trades on site and the amount of transport associated with the construction. A concrete frame has better thermal and acoustic properties than a steel frame, thus reducing the demand for thermal and acoustic insulation. However, concrete columns take up more space than an equivalent steel column. Thus, to maintain the position and overall design of the windows, the columns would need to be positioned further into the building. This had implications on the proposed furniture layouts. It also increased the cantilever needed for the external balconies, and, thus, the thickness of the concrete floor needed to achieve this cantilever. This thicker floor would result in a deeper floor zone and, thus a lower floor to ceiling height.…	*Becoming low maintenance* *Becoming simple of the design* *Becoming simple of the construction process/ programme* *Becoming thermally efficient* *Becoming acoustic* *Becoming less functional* *A new becoming of the facade design*
This had a knock-on effect on the space planning of the apartment corridors, the position of the lift and the layout of post boxes at ground level.	*A new becoming of the layout*

Figure 2.4 Extract of a design sequence marked-up using Deleuze's ontological concepts

In addition, this figure reveals a broader becoming: 'becoming sustainable' that was developed as a result of these experiments. As I noted in Chapter 1, 'becoming low maintenance of the building' implies greater durability and less building work across its life. A 'becoming simple of the design' implies less bespoke materials manufactured individually, and fewer specialised trades travelling greater distances. 'Becoming simple of the construction process/programme' implies less time on site, less transportation and reduction in road closures and so forth. All of these implications can be understood as different ways of thinking about sustainability: different contributions to a broader 'becoming sustainable' of the building.

The singularity and the threshold are concepts that only partly explain how becomings come together into blocks. The other aspect driving this process is reflected in Deleuze's proposals for force relations and the concept of the machine. Patton captures the concept of 'force' effectively in the following description:

> 'Force' should be understood, in abstraction from any determinate kind of action or interaction, to encompass all of the means by which bodies interact with one another. In this sense, 'force' is equivalent to 'power' in its primary capacity to do or to be certain things. Forces are the potentials for acting and being acted upon.
>
> (Patton, 2000: 52)

In this quotation, Patton suggests that as well as a long-term tendency and a set of thresholds, all becomings contain a 'capacity to affect and to be affected'. It is the combination and relation between these capacities that determine why one becoming might be favoured over another – i.e. why one becoming might be dominant over another. As with the other concepts discussed above, this concept of force can be seen in the example used in Chapter 1.

The becoming corrosive of salt cannot be separated from its capacity to affect specific entities that have a capacity to *be* affected – i.e. steel. Similarly, the introduction of new elements into the assemblage like specialist paint and the crossing of thresholds in the development of that assemblage (the shift from steel to concrete) are directed by this 'field of force relations' (Patton, 2000: 55). With this in mind, one might conclude that in order to understand the broader 'becoming sustainable' of a developing assemblage, it is important to appreciate why entities are introduced or removed from an assemblage, and to do this by thinking about what they 'do' within this assemblage: their capacity to affect and be affected.

The machine

This focus on what an entity 'does' is pivotal to the last concept I would like to draw from Deleuze's philosophy: the machine. The concept of the machine was first introduced in Felix Guattari's essay 'Concrete machines' (1984) and later became integrated into Deleuze's ontological framework following their collaboration on the two books that form *Capitalism and Schizophrenia* (2004; 2004a). The Deleuzoguattarian machine serves two important functions in the process of

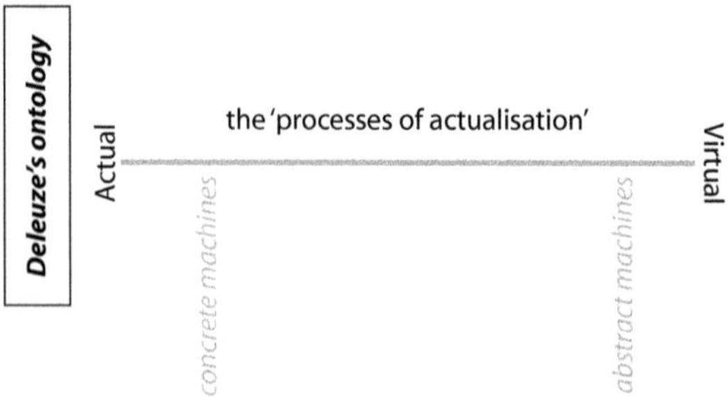

Figure 2.5 Machines and the actual/virtual

actualisation. The first function can be understood as a kind of filter or sieve. This filter is used to determine which becomings will progress towards the actualisation of a new entity and which will be blocked. The second role of the machine is to bind these becomings together to form the entities (assemblages) outlined above.

To explain this two-part process, DeLanda draws an analogy with the process of rock formation (1997: 59–60). When one looks at rock strata, he notes, one is able to identify strata formed from similarly sized pebbles. This homogeneity occurs because the river used to transport such pebbles acts as a 'sorting machine' filtering out certain kinds of rocks through the speed of the currents driving the movement of the water, the temperature of the water and interaction with the clay deposits within the river.

This sieve-like process is then coupled with a binding operation in which individual pebbles are combined to form a much larger entity: sedentary rock. As DeLanda notes, this 'cementing' process is made possible through the dissolved substances within the water that percolate through the pebbles and crystalise to form a solid mass. The result of this two-part process is a 'double articulation': two operations that transform entities at two different scales – i.e. that transform the pebbles within the river and transform the landscape through the creation of new rock formations.

For DeLanda, this example helps to explain two kinds of machines within Deleuze's philosophy: concrete machines and abstract machines. In this example, the river acts as a concrete machine whilst the broader process of stratification acts as an abstract machine. As the terms suggest, the difference between them lies in their position on the line between the actual and the virtual (Figure 2.5).

The above diagram shows how the abstract machine and the concrete machine are positioned on opposing sides of this actual/virtual line. This makes sense when one thinks that a river is an identifiable, *concrete* entity that acts upon other identifiable, actualised entities. Stratification is far more abstract in nature: a conceptual diagram that acts on many different entities, hence the reason for speaking about the stratification of rocks *and* the stratification of social strata for example.

DeLanda's analogy shows us that abstract machines determine how concrete machine operate, which, in turn, determine how new entities are formed. The result is, what Abrahams and Hillier term, a 'cascade of machines' (2016). Yet, this cascade does not always flow from the abstract to the concrete, but, as Bryant argues, this cascade can flow between and across different machines operating at different stages of development (2014).

These observations can be illustrated using the example of environmental assessments used to introduce this book. Across the last two chapters, I have shown how the abstract machine of essentialism has been used to form the diagram of the modelled concept, which has, in turn, been used to set out concrete machines such as the Code for Sustainable Homes and BREEAM assessments. This cascade of machines has, as Bryant notes, affected the way other machines operate. For example, the sustainable assessment tool as concrete machine affected how the designer as concrete machine selected entities to be included in the developing building assemblage. This, in turn, affects how the building as concrete machine affects how future residents behave and the furniture they select for their homes.

Making Deleuze useful to planning assessments

This short review of Deleuze's concepts and their relationship with essentialist assessment tools and the design process is not intended as a way of 'proving' that Deleuze's philosophy is 'right' or that it is better than any other theoretical framework. Rather, it is intended to reveal two things: first, that Deleuze's concepts can be understood using 'real-life' examples, and, second, that it offers opportunities for responding to the issues presented in Chapters 1 and 2.

In terms of the latter, the analytical exercise above shows how Deleuze's concepts help to explain the different understandings of sustainability that arise from the design process (understandings currently ignored in formal, essentialist assessment methods). It also shows how Deleuze's concepts capture some of the points highlighted in non-essentialist planning literature. These include Fainstein's suggestion that planning should not ignore the broader forces (broader blocks of becoming) acting on the development process (2000; 2005), and Stein and Harper's suggestion that meaning, and the way one assesses meaning, should respond to the unique developments undertaken *during* the design or planning process, in this case, *during* the design process (Harper and Stein, 2006; Stein and Harper, 2012). Likewise, it reflects many of the so-called 'post-modern' calls for more flexible and context-sensitive assessments designed to reflect 'real' decision making (Alexander, 1998; Borri, 1998; Fischer, 2003).

This chapter makes a case for using Deleuze's concepts to develop an assessment tool that responds to concerns raised by actors working in the design and development process, academics working in the field of assessment methodology and planning theorists. In the next chapter I will consider what is needed to make this transition: i.e. how one operationalises these concepts to make them useful to concerns and issues raised in the spatial disciplines.

Notes

1 Hence, why the sum of these nine categories form 100% of the assessment.
2 I use the term 'plan-making' to highlight a broader set of processes including policy-making, strategic plan-making, local plan-making, and urban and building plan-making.
3 The strategic use of this essentialist/non-essentialist frame is discussed further in Chapter 9.
4 A working definition of these concepts is outlined below.

Part 2

How to make Deleuze useful

In Part 1 I made the case for using Deleuze's concepts to form an alternative to essentialist assessment tools. In Part 2 I consider how one goes about making these concepts useful by exploring some of the attempts made by other spatial theorists.

3 Attempts to make Deleuze useful

Introduction

In the last chapter I introduced a set of problems associated with essentialism and the kind of modelled concept-based tools derived from essentialism. I discussed a number of philosophical concepts from Deleuze's philosophy and demonstrated how these concepts can be used to explain the design process and how they might be used to set out a new, non-essentialist method for assessing such designs.

Aim and structure of the chapter

The aims of this chapter are two-fold. First, it is intended to provide further clarity on the Deleuzian concepts discussed in the last chapter by showing how they are used in different fields. Second, it aims to identify a number of lessons from other Deleuze-inspired spatial theorists about how one might translate these concepts into a theoretical framework that can usefully contribute to a specific line of enquiry. These aims reflect the principle objectives of this book – namely, to translate Deleuze's concepts into a practicable planning tool (see Chapter 1).

To achieve these aims, the following text reviews and critically discusses a selection of Deleuze-inspired studies in detail. Such detail is important if one is to form meaningful conclusions about translating and adapting a Deleuzian concept from philosophy into planning theory and planning practice. With this in mind, I have centred my review on three important studies. These studies act as an organising device to bring in other cognate debates and points.

Selecting the three Deleuze-inspired voices

Over the last few years there has been a growing community of spatial theorists whose work makes direct or indirect references to Deleuze's concepts. Leafing through a range of spatial theory journals, one can quickly spot passing references to 'assemblages', 'becomings' and multiplicities'. To choose three studies from this growing body of knowledge, I have applied three selection criteria.

Criterion 1: analysis and engagement

As noted above, the broad objective of this book is to consider whether Deleuze's philosophy can be translated into a non-essentialist assessment that can be operationalised in planning and development practice. This objective suggests using Deleuze in a way that is not based on analysing existing forms of assessment and planning practice alone but also offers new tools for engaging in such practices. This is what the planning theorist John Forester termed 'reconstructive' as well as 'deconstructive' planning theory (Forester, 2007). With this in mind, my first selection criterion will include one Deleuze-inspired study from each of the following three groups:

1 using Deleuze's philosophy as a form of analysis;
2 using Deleuze's philosophy as a form of analysis and as a method of engagement in space/spatial issues;
3 using Deleuze's philosophy predominantly as a method of engagement.

In doing so, I hope to gain some insight into the way Deleuze has and has not been useful to inform different research objectives.

Criterion 2: ontological concepts

As I noted previously, the concepts I have selected from Deleuze's philosophy capture his broader ontological understanding of 'the real' (see Chapter 2): formed from the actual, the processes of actualisation and the virtual (Figure 2.3). Whilst spatial theorists and practitioners have been keen to draw on Deleuze's concepts, there have been fewer attempts to consider these concepts as part of a broader ontological framework (DeLanda, 2002). With this in mind, the second selection criterion was used to select studies that draw on more than one Deleuzian concept and are supported by attempts to understand the ontological setting in which these concepts were created.

Criterion 3: representativeness

It is important to ensure that these interpretations and their subsequent application reflect current thinking within the spatial field of study. With this in mind, I have included studies whose contribution to the field of knowledge is supported by other Deleuzian-inspired spatial scholars and non-Deleuzian scholars alike.

The following studies effectively met all three selection criteria as detailed in the introductory section of my review for each study:

1 Mark Bonta's study of changing land use in Olancho, Honduras (Bonta, 2001; Bonta and Protevi, 2004);
2 Mark Halsey's study of, and proposals in response to, textual violence in the Goolengook forest in Australia (Halsey, 2006);

3 Jean Hillier's proposals for a multiplanar theory and method in spatial planning (Hillier, 2007).

As part of this selection process, a number of other studies were considered and dismissed. These include Nigel Thrift's proposals for a 'Non-representational Theory' in geography (Thrift, 2007) and DeLanda's proposals for 'A New Philosophy of Society' based on his study of spatial assemblages (2006). These two studies were not selected for different reasons. Thrift's study demonstrates a strong understanding of Deleuze's concepts but does not explore Deleuze's ontological framework to the same extent as the geographical study included in the selection: Bonta and Protevi's study of 'geophilosophy' (2004). As demonstrated in Chapter 2, DeLanda's study is part of a larger framing of Deleuze's ontology. However, as a philosophical study undertaken within the spatial disciplines, DeLanda's proposals, I feared, might not fully represent the issues and concerns specific to the spatial disciplines, particularly when compared to the three spatial studies selected.

The three studies that were selected (Bonta and Protevi; Halsey; Hillier) each focus on a problem within their respective field and develop Deleuze's concepts to respond to that problem. To help me identify useful lessons from these studies, I have broken each detailed review into four sub-sections:

1 brief introduction to the study;
2 the problem directing the study;
3 the core concepts drawn from Deleuze's philosophy;
4 the resulting theoretical framework and its gaps.

Bringing these three studies together, the 'Discussion' section of this chapter will consider what lessons can be learned from all three studies, their individual benefits and their failings and how they might help direct my own use of Deleuze's philosophical concepts introduced in Chapter 2 and developed in detail in Chapter 6.

Bonta: analysing 'complex spaces'

The first voice in my review belongs to the geographer Mark Bonta, as first captured in his analytical study of changing land-use in the Olancho region of Honduras, South America (2001), and developed in his subsequent text with John Protevi (Bonta and Protevi, 2004).

A brief introduction to Bonta's study

Mark Bonta's study and his subsequent work with the philosopher John Protevi situates him as 'among the first to bring Deleuze and Guattari together with complexity theory . . . [to respond to some of the] questions facing geography' (Bonta and Protevi, 2004: 7). These efforts provide us with one of the most thorough and detailed readings of Deleuze's philosophy and one of the only attempts to translate these philosophical concepts into a glossary intended for spatial theorists. Bonta and Protevi refer to this as a 'geophilosophy' (DeLanda et al., 2003; Bonta and Protevi, 2004).

Bonta uses this geophilosophy to make sense of his ethnographic observations of land use in the Olancho region of Honduras, South America. Deleuze's philosophy and his concept of the *assemblage* in particular help Bonta analyse and explain complex combinations of people, objects and ideas.

This geophilosophy has been used as support in a number of subsequent Deleuze-inspired studies. Such studies include proposals for a Deleuzian 'understanding [of] the entangled relationships between globalization, climate change, capitalism and indigenous peoples' (Cupples, 2011: 10), proposals for new ways to understand geographic scale (Marston et al., 2005), the development of households in Quintana Roo, Mexico (Normark, 2009), and developments in 'relational' geography through advances in complexity science (Jones, 2010). It has also influenced proposals for new ways of engaging in spatial planning (Hillier, 2007) and politics (Palmas, 2008).

Despite its influence, there have been few attempts to understand how Bonta expands and (re)creates Deleuze's concepts to make them useful to his particular field of enquiry, how these concepts contribute to Bonta's theoretical framework and what limitations and problems result from these concepts. This part of the study helps fill these gaps in the literature.

The problem directing Bonta's study

Bonta is attracted to Deleuze's philosophy after rejecting many of the methods used by geographers to understand and analyse the complex world around us. In his collaboration with the philosopher John Protevi, Bonta discusses and rejects three existing approaches to spatial analysis. Each of these approaches, they note, shares a common focus on 'the ideal of description and explanation' (Bonta and Protevi, 2004: 200). Bonta and Protevi's critique is poorly supported with examples from comparative studies undertaken in Honduras. To rectify this shortcoming, I will discuss and, where possible, illustrate these three approaches by identifying and using land-use studies undertaken in and around Bonta's study area.

'Molar' analysis

In the first analytical approach rejected by Bonta and Protevi, an area is studied in reference to one or several broad categories. There have been a number of studies undertaken over the last twenty years that might reflect this 'molar' approach to analysis. Such examples might include attempts to relate deforestation to factors such as foreign debt (Gullison and Losos, 1993), tenure security and the leisure preferences of local residents (Godoy et al., 1998); time spent in early, formal education (Godoy et al., 1998a); population and the Human Development Index (HDI) (Jha and Bawa, 1999); population density (Pfeffer et al., 2005); and the interaction between different types of land use (Ruf and Schroth, 2004). These examples from land-use literature help us appreciate Bonta and Protevi's argument that

> far too many researchers . . . have burdened us with the 'molar' rain forest, biome, life zone, frontier, town, city, *campesino.* These are indeed molarising – normalising – processes at work that ensure we will encounter similarity and linkages wherever we go.
>
> (Bonta and Protevi, 2004: 42)

This quotation captures two criticisms against this approach to analysis. First, it questions the validity of findings that relate one type of land use to one or several influential factors. Second, it suggests that such studies have negative effects on the way these spaces operate. It does this, Bonta and Protevi suggest, by fixing our understanding of entities found in these spaces and the relationships between them.

Analytical modelling

In many ways, these problems are overcome in attempts to develop analytical models. Rather than focus on one or several factors, these attempts consider the joint contribution of social, economic and environmental factors in changing land-use.

This modelling approach has gained a great deal of attention over the last fifteen years. Many of the models related to Honduras have focused on deforestation. National Forest Monitoring and Assessment (NFMA) and the Global Forest Resources Assessment (FRA), for example, both map forest types and model their changes over time (Food and Agriculture Organisation, 2010). Other integrated models, like the Land Use and Cover Change (LUCC) models, have taken a broader view. These integrated models look at a combination of influential factors affecting different kinds of land use change in different parts of the world (Pauleit et al., 2005).

Despite their growing popularity, Bonta and Protevi argue that this second area of analysis is problematic for two reasons. The first concerns the impossibility of modelling the complexity seen in complex social systems:

> we are still far away from being able to model social systems successfully. The biggest problem . . . is (that such models are not) very useful for those committed to a realist attempt to capture features of the world rather than merely modelling phenomena.
>
> (Bonta and Protevi, 2004: 33)

Modelling, they suggest, typically identifies a phenomenon but fails to explain how this phenomenon comes into being. This argument is supported by others working in the field of land-use planning (Nagendra et al., 2004; Monroe and Muller, 2007).

For Bonta and Protevi, the second problem with these models relates to their use in practice. For Bonta and Protevi,

> [t]he problem comes when people write about (neoliberal) economics as if they were only a matter of assumptions and models rather than prods for concerted efforts to produce a social reality conforming to the model's assumptions.
>
> (Bonta and Protevi, 2004: 199)

As with their critique of 'molar analysis', Bonta and Protevi believe that the models used to explain complexity fail to capture many of the subtle changes and relationships within practice. These models, they argue, become blueprints for our understanding and subsequent efforts to engage with the complex world around us.

This argument against modelling draws links to my own research agenda. In Chapter 2, I outlined attempts to model abstract concepts like the 'sustainable home' and the 'sustainable, newly constructed building'. And like Bonta and Protevi, I questioned the idea that such 'modelled concepts' should be used to change the way urban and building designers understand and engage in complex spatial problems.

Thick description

Geertz's 'thick descriptive' methods have had a strong influence in ethnographic research over the last forty years (Geertz, 1973). In the region of Honduras in particular, Dean offers us a thick descriptive account of family life in Northern Honduras (Dean, 1988). More recently, Pine offers a highly descriptive account of one Honduras citizen and his movements across different spaces (Pine, 2008). For Bonta and Protevi, 'thick descriptions' do not account for the

> irreducibility of distributed spatiotemporal networks of embodied artisans in 'resolving' complex problems by real life operations.
>
> (Bonta and Protevi, 2004: 25)

For Bonta and Protevi, therefore, these three forms of analysis fail to understand and explain the complex factors operating in these spaces. Taking this further, they also fail to explain how the things we see around us came into being through 'real life operations' (Bonta, 2001; Bonta and Protevi, 2004).

Of course, these three approaches criticised by Bonta and Protevi are not unique to geography. Planning literature includes numerous studies that Bonta and Protevi might term 'molar explanations', such as the attempts made to explain changes in the economic growth of a region or country through developing Information and Communication Technology (ICT) infrastructure (Roller and Waverman, 2001; Sridhar and Sridhar, 2007, Czernich et al., 2011). Planning literature also outlines numerous attempts to model a range of complex, inter-related factors (Batty, 1996; 2011; 2011a; 2011b) as well as efforts to understand complexity through thick description (Flyvjberg, 2001; Maginn, 2007). This suggests that Bonta and Protevi's arguments might be used to challenge studies undertaken in other spatial disciplines. In terms of this study, Bonta's critique of these methods, and of molar analysis in particular, shares some of my concerns with the limited approach to assessment seen in formal, essentialist assessments like the Code for Sustainable Homes and BREEAM (Chapter 1 and 2).

The core concepts drawn from Deleuze's philosophy

To resolve these problems, and to outline an alternative method of analysis, Bonta draws on several concepts from Deleuze's philosophy.

The assemblage and complex spaces

One of the most important Deleuzian concepts in Bonta's study is that of the assemblage and his expansion of this concept to form the concept of 'complex space'. For Bonta, these concepts capture his observations from the field. Drawing from his field notes, Bonta describes the way that

> [c]offee farms were being taken over by cattle; beans were taking over forests; forests were taking over ranches; Hurricane Mitch had stripped away cattle pastures, beanfields, coffee, and forests alike.
>
> (Bonta and Protevi, 2004: 172)

This example shows that a piece of land being used as a 'coffee farm space' was also becoming a 'cattle farm space', which was, itself, becoming a 'forest space' and so forth. The result is a complex combination of different uses all operating on the same piece of land and all in a state of transition. Bonta describes these combinations as 'complex spaces'. Rather than trying to model or capture these 'complex spaces', Bonta argues that we must understand them as a combination, or an 'entanglement' of different spaces. And, thus, when we analyse them, Bonta suggests we must study the way a coffee farm space becomes entangled with a cattle ranch space and a forest space (Bonta, 2001; Bonta and Protevi, 2004).

Bonta observed:

> If one were 'plugged into'[1] the cattle-ranch (space), one was to a large extent predetermined and at the very least codetermined by a complex system quite different to that of one's neighbour, who was plugged into the complex system of coffee farming.
>
> (Bonta, 2001; Bonta and Protevi, 2004: 172)

This example shows how a coffee farmer would be part of a set of processes particular to planting, growing, harvesting and preparing coffee beans. These would be very different to the processes required of his neighbour, a cattle farmer. Likewise, these two sets of processes would require similar material and non-material entities but adapted in different ways, such as varying degrees of education, population density and machinery. They would also demand differing quantities of land at differing times (Bonta, 2001; Bonta and Protevi, 2004).

To add to the complexity, Bonta observed that entities operated in more than one space. Bonta noted that, as a coffee space was becoming[2] a cattle space, the coffee farmer was also becoming a cattle farmer:

> Because individual human bodies are the nexus for myriad forces that traverse them, they are easily put to work for different complex systems at the same time . . . [resulting in] rancher-conservationist-teachers, peasant farmer-ranchers and rancher-logger-coffee growers.
>
> (Bonta, 2001; Bonta and Protevi, 2004: 172)

This short description highlights the range of roles undertaken by different members of the community and why traditional categories fail to capture this complexity.

If we extend this to non-human entities as Bonta and Protevi suggest, we might consider water on the land as simultaneously cleaning the coffee, cooling the coffee machines, cleaning and hydrating cattle, feeding the grass that feed the cattle, transporting and nourishing plant seeds from the forest, directing the flow of tree roots and so forth. In the same way that Bonta asks us to think of a 'complex space' as a combination of other spaces, Bonta asks that we think of these spaces as a combination, or 'assemblage', of human and non-human entities. This illustration of complex space as an assemblage formed from other assemblages captures Deleuze's argument for 'the relations of exteriority' because it suggests that each assemblage can be understood as an individual in its own right as well as an entity operating as part of a broader assemblage – i.e. the complex space. Bonta's approach to assemblages, therefore, shows parallels with the illustration I present in Chapter 2.

Like DeLanda and Patton, Bonta argues that human and non-human entities within an assemblage should be understood according to their capacities to affect other entities: their 'power' (Patton, 2000; DeLanda, 2006). Thus, in the above example, water has the capacity to cleanse, cool, hydrate, transport and attract other things in a range of assemblages. These capacities allow water, for example, to 'plug into' different processes in the same assemblage, or 'plug into' processes across other assemblages (Bonta and Protevi, 2004). For Bonta, this conception of the assemblage and its relation with other assemblages reveal the entanglements that characterise 'complex spaces' (Bonta, 2001; Bonta and Protevi, 2004). Mapping these assemblages over time, he argues, should help us understand the dynamism of land-use (Bonta and Protevi, 2004).[3]

This last idea suggests that we can understand complexity by mapping the assemblages we identify. This idea is not new to geography or planning. Similar attempts at mapping assemblages are often associated with Actor Network Theory (ANT) and its methodological nostrum that one should follow the actors, be they human or non-human, and map the networks that they construct. ANT has a long history in the social sciences, making these methods accessible to both geographers and planners (Murdoch, 1995; 1997; 1998; 2001; Farias and Bender, 2010).

This growing popularity in ANT and the idea of mapping assemblages has led to a number of criticisms. For Allen, these maps help us to appreciate how heterogeneous elements can work as a collective without resorting to one universal explanation. However, whilst this may prove beneficial to some urban geographers, he suggests that it may result in 'endless description' and 'weak conceptualisation' (Allen, 2011). Robbins and Marks also draw attention to these two risks. For Robbins and Marks, many efforts to map assemblages to date have done so by describing rather than conceptualising relations. When these assemblages have extended beyond description, they have tended to critique existing concepts rather than offer us new ones (Robbins and Marks, 2010). We can identify similar criticisms in other areas of geography. In their review of efforts to map institutional assemblages, Whittle and Spicer conclude that most studies did not offer a useful, workable theory (Whittle and Spicer, 2008).

Rather than reject Bonta's use of these concepts, these criticisms demonstrate an important distinction between Latourian (ANT) and Deleuzian assemblages. For Latour, the assemblage concept provides a means to map 'networks' of material and non-material entities. Deleuze, however, places greater emphasis on the assemblage as an ontological concept relating to other ontological concepts in his theoretical framework. Mapping assemblages helps the researchers understand what drives their creation, or what Deleuze terms 'becomings'. It also helps the researchers understand the potentials for re-creating them differently in the future, or what Deleuze termed an assemblage's 'virtual diagram', or 'multiplicity' (see Chapter 2). We might argue, as Allen does, that compared to Deleuze's approach, ANT places too much emphasis on capture and description. Indeed, if we think back to Bonta's critique of other analytical approaches, Bonta might also reject an ANT understanding of assemblages on similar grounds.

Smooth and striated space

Above I have shown how Bonta's observations helped him identify some of the components that form a coffee farm assemblage and a cattle farm assemblage. But to explain how these components organise space, one must decide what spatiality can be attributed to these assemblages.

Traditional ways of understanding space and the organisation of space are characterised by three dimensions: things in the real world, representations of this world and the subject who forms the representation (Buchanan and Lambert, 2004; see also DeLanda, 2006). These distinctions are inappropriate for studying assemblages because many different entities contribute to the way assemblages are formed and the way they operate (Bonta and Protevi, 2004; DeLanda, 2006). Nigel Thrift has similarly argued against the use of representations in geography. For Thrift, these representations limit our ability to engage or understand the way different elements and forces 'perform together' (Thrift, 2007).

To avoid these traditional classifications and the problems associated with them, Bonta calls on Deleuze and Guattari's concepts of 'striating' and 'smoothing'. In *A Thousand Plateaus*, Deleuze and Guattari note that

> there is a significant difference between the spaces: sedentary space is striated, by walls, enclosures, and roads between enclosures, while nomad space is smooth, marked only by 'traits' that are effaced and displaced within the trajectory.
>
> (Deleuze and Guattari, 2004b: 420)

When Deleuze talks about smooth space, therefore, he talks about spaces that are understood as part of movement. As we move through smooth space, we remove the characteristics identified by others. In contrast, striated space is structured, enclosed and defined. These structures determine how we interact with things and, to some extent, how we move through space. From this passage, it is evident that Deleuze does not intend the concepts of 'smooth' and 'striated' to be properties of a space. Rather, he uses them to explain ways of making or adapting space.

As I noted in Chapter 2, whilst these concepts are oppositional, acts of smoothing and striating are not separate and exclusive. Smoothing and striating space occur together and as part of the same set of processes (Deleuze and Guattari, 2004b: 427). In the following, I will show how Bonta draws on these concepts to explain how different assemblages simultaneously organise (striate) and disorganise (smooth) space to meet their unique demands. Whilst considered in terms of land use rather than design and regulation, the insight offered by Bonta provides some important clues about the role played by different assessments and how they relate to the way assemblages (such as the designer assemblage or the regulator assemblage) organise and disorganise space.

Drawing on Deleuze's concept of the assemblage, Bonta provides us with a four-hundred-year history of assemblages operating within the region's 'complex spaces'. These assemblages are not limited to definable land-uses like the coffee farm assemblage but also include more abstract assemblages like 'the state assemblage', 'the church assemblage' and 'the environmental assemblage'. By mapping this history, Bonta shows how assemblages are formed, re-formed, taken-over and adapted by striating and smoothing space. A review of this history reveals a number of key points. In the following, I will discuss these points in the context of Bonta's study before reflecting on their relevance to my own line of enquiry concerning the use of modelled concepts in assessment design.

The first point I would like to discuss concerns acts of 'smoothing'. Bonta's history highlights the fact that spaces are never empty. When an assemblage seeks to striate land to meet its needs, it must engage with the striations instated by other assemblages. Bonta identifies instances when individuals within an assemblage have removed existing striations through 'smoothing'. Bonta explains how, in the 1500s, colonial settlers introduced cattle to the existing agricultural assemblages used to feed the existing Indian populations. These cattle spread over the plains eating and trampling the vegetation. In doing so, they 'smoothed' striated planting fields (Bonta and Protevi, 2004: 176).

Such attempts at smoothing the striations set out by existing assemblages have not always resulted in an effective 'take-over'. Bonta identifies instances when

> [r]anchers deploy[ed] cattle to effect takeovers of other territories, for example the deterritorialisation [smoothing] of coffee space [and] . . . so coffee space had to be fenced and gated.
>
> (Bonta and Protevi, 2004: 177)

In this example, the smoothing effects of cattle were resisted. The result was a more heavily striated space made possible by expanding the assemblage to include new entities (fences). This example shows how an assemblage might expand and re-form through its interaction with (or resistance to) other assemblages operating in the same space.

The effects of this expansion are relatively limited. The introduction of fences into the coffee farm assemblage only affected the assemblages already operating in the same space. Yet this is not always the case. In a different example, Bonta explains how the conservation assemblage was expanded in the wake of an event: Hurricane Mitch. This event triggered attempts to striate space according to the demands of a new addition in the assemblage: buffer zone concepts.

> [These] buffer zone concepts [were used to] . . . promote a zone of contact between land-using humans and Nature . . . while making the forest space itself off-limits to all localised land-uses, but useful as a pure reserve of water and biodiversity, and accessible only to ecotourists and scientists.
>
> (Bonta and Protevi, 2004: 184)

By creating and including buffer-zone concepts, this example shows how the conservation assemblage divided human and non-human uses. The effects of which were to promote the interaction of select assemblages and prevent the interaction of others.

These examples highlight two important points. First, they show that all entities, human and non-human, within an assemblage are motivated to striate and smooth space to meet their unique operational demands. This is equally applicable to the actors in the design and regulatory process who organise space to meet the demand of actors in their assemblage. This also illustrates the distinction discussed above between a Deleuzian and an Actor Network Theory approach to assemblages and mapping. In his review of Callon's ANT study (Callon, 1986), Murdoch notes:

> Although we might legitimately wish to ask, for instance, what motivates entities such as the scallop scientists to become actors, ANT theorists are unwilling to enter into a discussion of motivations, intentions, and the like, as these smack too much of some internal, already solidified notion of social

> agency. . . . Motivations are thereby discussed in only the most general and abstract terms.
>
> (Murdoch, 1998: 746)

This same criticism could not be applied to Bonta's discussion on 'complex spaces' and the relationship between coffee farm and cattle farm assemblages in particular.

Second, these two examples show a distinction between two kinds of entity. This distinction relates to the way these entities are introduced into a complex space and how they affect assemblages within a complex space. This distinction is captured in the way Bonta negotiates two Deleuzian concepts: the multiplicity and the referential model. As I will show below, this distinction in Bonta's study offers parallels with the distinction between essentialist and non-essentialist assessments discussed in Chapters 1 and 2.

Multiplicities and referential models

Whilst roaming cattle had a material effect on the coffee farm assemblage, they were not responsible for deciding how the coffee farm coped with this effect – i.e. by adding fences into the assemblage. The decision to expand the coffee farm assemblage to include the fence came from *within* the coffee farm assemblage itself: by the operational and spatial demands of coffee plants, harvesting and sowing machines. Likewise, the introduction of the fencing would have material effects on the cattle farm. But again, the coffee farm assemblage would not decide *how* the cattle farm assemblage responded to the effects of the fence. Any entities added to or omitted from the cattle farm assemblage to respond to this fence would have been determined from *within* the cattle farm assemblage.

Unlike the fence, the buffer-zone concept did not originate from the unique demands of entities operating within the assemblage. Bonta informs us that this concept was 'based on the US model . . . the "perfect" park out of the US textbook' (Bonta and Protevi, 2004: 183–184). It was, we might argue, a model created within a different context, in a different country, at a different time and with a very different purpose.

Additionally, the buffer zone (or model) was more intrusive than the fence. This model sought to change the way these assemblages operated, by 'persuading land-users to adopt, "land-friendly practices" ' (Bonta and Protevi, 2004: 184). And in doing so, they sought to determine *how other* assemblages organised themselves by the entities they selected or omitted from the assemblage.

Bonta's study provides two different kinds of entities, and, thus, two ways of creating/re-creating an assemblage. The first kind of entity (like the fence) is introduced into an assemblage to meet the operational demands of other entities within that assemblage, and to respond to the effects of 'entanglement'. Thinking back to the example I give in Chapter 1 and expanded in Chapter 2, one might argue that this reflects the assessments made during the design process: drawing from what Deleuze termed an assemblage's 'multiplicity' (or 'virtual diagram').

The second kind of entity, like the buffer-zone concept in Bonta's study or formal assessments such as the Code for Sustainable Homes or BREEAM in this study, is developed in a different context and serves a different purpose. Such entities are introduced into a complex space to change the way other assemblages operate. An assemblage is re-created not by drawing on the assemblage's multiplicity but by reference to a pre-defined model created in and for a different assemblage. Taken together these two kinds of entity provide us with a working example of Deleuze's multiplicity and its distinction from the referential model.

This distinction is an important one, not least because it helps explain the kind of spatial problems that Bonta believes are best suited to a Deleuze-inspired approach. For Bonta, the spaces he observed in Olancho are complex. They are formed from different assemblages all operating together resulting in their 'entanglement'. Whilst these assemblages have material effects on each other, this does not change the way they operate. For Bonta, each assemblage operates in its own way, adding or removing entities to reflect these operational demands. As they do so, assemblages must smooth the striations set out by others and striate the land to meet their own requirements. This kind of space, he argues, lends itself to a Deleuze-inspired approach. Deleuze, he argues, presents us with concepts that can be usefully incorporated into a theoretical framework used to make sense of, and explain, the way different assemblages come into being in the absence of a dominant, causal agent – i.e. emergence. The theoretical framework captured in Bonta's study is captured in Table 3.1 below.

Table 3.1 Mark Bonta's theoretical framework (Bonta, 2001; Bonta and Protevi, 2004)

Deleuzian concepts	*Bonta's concepts*
Assemblage: A combination of human and non-human entities operating in their own way as part of the assemblage.	***Complex space:*** A complex space is formed by a combination of assemblages.
Smooth and striated: Each entity striates and smoothes a given space to meet its unique operational demands.	***Complex space:*** A complex space can only exist when a given space is not dominated by any one set of striations defined by one or several assemblages.
Becomings: Becomings are potentials for an entity to operate in certain ways.	
Multiplicity: A multiplicity is the combination of all potentials in all the entities in an assemblage. It provides the conditions for developing an assemblage.	
Referential model: The referential model is used to determine which potentials will and will not be operationalised.	

The resulting theoretical framework and its gaps

The left-hand-side column lists five of Deleuze's key ontological concepts and the way Bonta interprets these concepts to make them useful to his study. Running from top to bottom of the column, these concepts are used to explain the combination of entities operating in Bonta's field of study (assemblages), how they organise space (striations and smoothness), how they are able to operate in this way (becomings), where these potentials come from (multiplicity) and one way in which this process is directed in policy assemblages (referential models). This left-hand column presents a very similar picture to the one outlined in Chapter 2, and my own attempt to explain these ontological concepts using examples from the design process.

The second column to the right-hand side identifies the concept that Bonta creates to meet the specific demands of his project. As the table suggests, Bonta develops the concepts of the 'assemblage' and 'striated and smooth space' to form the concept of 'complex space'.

This concept of 'complex space' is a strong addition to the concepts offered in Deleuze's seminal texts. However, it also leads me onto what I believe to be the principle gap in Bonta's theoretical framework. According to Bonta, not all spaces can be considered as complex spaces. For Bonta, spaces cannot be considered complex if one or several assemblages dominate land-use. These dominant assemblages, he argues, have direct, material effects on other assemblages. They determine how other assemblages operate, how they add or remove entities and how they organise space. Rather than explaining these relations, Bonta holds that Deleuze's theoretical framework provides us with the tools we need to problematise the dominant position held by such assemblages (Bonta and Protevi, 2004).

This begs the question, how does this relate to other studies outside of Olancho? For Bonta and Protevi, complex spaces can be found in developing rather than developed countries where the state assemblage has not assumed a dominant role over other assemblages (Bonta and Protevi, 2004). However, for me, this final distinction is too simple. Do complex spaces not exist in Western, developed countries? If we sit at any urban square in the UK, can we not see a number of entangled assemblages operating in the area? After all, the entities operating in these areas are not all part of a state assemblage. Drug sellers, homeless people, stray cats, pigeons, children playing with bottle tops and so forth also operate *within* other assemblages not defined by a state assemblage. And each of these assemblages smooth and striate space to meet their own unique demands.

Additionally, I question whether 'the state' can be understood as a single, monolithic assemblage. Like any other large assemblage, it too is formed from other assemblages that operate in different ways and seek to striate and smooth physical, as well as non-physical (such as policy), space to meet these unique demands. If we think back to Chapter 2, there is no reason to assume that the state assemblage should be governed by 'relations of interiority', whilst all other assemblages are governed by 'relations of exteriority'. It also assumes that *the* state assemblage implements its goals with clear and effective force and no discretion on the part of its subjects. With this in mind, it seems inappropriate to talk about *a* state assemblage as a unified body operating and striating space in a coherent way (Hillier and Abrahams, 2013).

I would like to suggest, therefore, that complex spaces are found in many different areas in many countries around the world. The design process I discussed in Chapter 2 could thus be described as a developing assemblage (the building assemblage) operating within a complex space. If we start with the assumption that all spaces are complex, then we are able to make fuller use of the ideas presented in Bonta's study and his later work with Protevi. We might then ask: what assemblages operate in the area? How do they entangle and affect each other through this entanglement? Which entities are introduced by other entities operating in an assemblage in complex space, and which are introduced as 'referential models' from other contexts? If we think back to Chapters 1 and 2, we see that modelled concepts would be a good example of this second type of entity. Do these models seek to change the way other assemblages are created and how they operate? And, if so, what are the effects on these assemblages operating in a complex space?

By adapting the concept of 'complex spaces', we can start to see how these kinds of questions might guide analytical methods in related disciplines such as spatial planning. This Deleuze-inspired understanding of complex space and its use alongside other Deleuzian concepts could be used to critique and draw lessons from studies of 'complex spaces' in ANT (Callon and Law, 2004). This could include an engagement with Law's proposals for 'baroque' and 'romantic' forms of complex space (Law, 2004) or Thrift's discussion on 'encountering' (Thrift, 2004).

Similar opportunities might exist in spatial planning literature more specifically. The concept of 'relational space', for example, has been developed by a number of influential planning theorists (Graham and Healey, 1999; Healey, 2006), as has its link to Deleuze's conception of space (Massey, 1999; Murdoch, 2006; Hillier, 2007; Malpas, 2012; Monno, 2012). Thus, it seems that an expanded understanding of Bonta's 'complex space' could help inform, revise and direct these theoretical developments in geography and planning. This expanded understanding of 'complex space' will be included in my own theoretical framework developed in Chapter 6.

Halsey: analysing and engaging in 'violence'

In my review of Bonta's 'complex spaces' above, I discussed the distinction he makes between two ways of creating an assemblage: by the assemblage's own 'multiplicity' (like the fence) or by reference to a model introduced into a different assemblage (like the buffer zone). Whilst Bonta presents a position against the latter, his study is mostly analytical. It explains the effects of the buffer zone as a reference model but does not consider how we might engage with its consequences. This is an important line of enquiry in this book because it informs the way we might respond to the modelled concepts used as the basis for formal assessments such as the Code for Sustainable Homes and BREEAM (see Chapter 2).

With this in mind, I turn now to the work of the ecologist and criminologist Mark Halsey. Like Bonta, Halsey draws on Deleuze's philosophy to consider the way assemblages develop as part of a 'complex space'. And, like Bonta, Halsey focuses on one rather than many complex spaces. However, by framing his work in the field of criminology, Halsey has a different reason for studying the way a complex space is developed over time. Halsey is interested in identifying and engaging with the consequences of naming and categorising (Halsey, 2006). These names are not too dissimilar to the buffer-zone concept drawn from Bonta's study or the modelled concept of 'the sustainable home' and the 'sustainable, newly constructed building' discussed in Chapter 2. For Halsey the consequences of naming should be conceptualised as a kind of 'violence' that affects the way assemblages develop over time (Halsey, 2006). By studying these effects (or 'violence'), he argues, criminologists can make judgements about the damages that might be incurred.

A brief introduction to Halsey's study

Since its publication, Halsey's call for an alternative approach to environmental criminology has been met with broad support (Brisman, 2008). For Gibbs et al., Halsey's work should be seen as part of a 'framework [that] will ultimately advance knowledge and practice regarding environmental crimes and risks' (Gibbs et al., 2010: 17). For Gough, this framework should be directed towards efforts to correct the violence of 'naming nature' (Gough, 2008). Gough describes these corrective measures as ways of 'un-naming nature' pursued by exploring a new becoming: 'becoming pedagogical' (Gough, 2008).

Whilst Gough uses Halsey's Deleuze-inspired proposals as a starting point for further development, others have drawn on Halsey's study to reveal a broader shift in ecological thinking. Fancy argues that Halsey's study demonstrates the benefits offered by a Deleuzian understanding of ecology (Fancy, 2011). This approach, he notes, draws on a line of thinking running through Deleuze's work related to geology and performance, which Fancy articulates as 'geoperformance' (Fancy, 2011). Halsey's interpretation of and use of Deleuze's concepts, therefore, have gained a great deal of support within his field of interest and have offered new opportunities for others to develop different lines of enquiry. But, like Bonta's study, there have been few attempts to understand *how* Halsey translates Deleuze's concepts to help him respond to this field.

The problem directing Halsey's study

Halsey's interest in Deleuze stems from the problems he identifies in five of the most popular theoretical perspectives used in the field of ecology: liberal ecology, ecomarxism, ecofeminism, deep ecology and social ecology. Taken as a whole,

Halsey argues that these 'modernist' approaches are unable to take account of environmental damage (Halsey, 2006: 34). This failing, he notes, arises from two problems.

The first problem relates to the idea that environmental damage has an identifiable 'cause'. This idea is fundamental to the five perspectives. For liberal ecologists, this cause can be traced to middle-class consumption patterns and the market's inability to meet these demands through more efficient products. For ecomarxists it is caused by the market itself and the class-system more broadly. Whilst for ecofeminists and social ecologists it is caused by the dominance of one group of society over another: men over women or, more broadly, the powerful over the less powerful. And, finally, deep ecologists argue that these problems are caused by a false distinction between humans and Nature (Halsey, 2006: 15–34). Whilst these arguments differ, they share a common 'tendency to write the "causes" of environmental problems in *monolithic* fashion' (Halsey, 2006: 35).

One might criticise Halsey's summary of these perspectives as overly simplistic. Looking across the literature from these five perspectives, one can identify work that bridges these distinctions. Mikulak, for example, shows how a Marxist analysis of labour captures the idea that humans cannot be understood in isolation from their environment, and, likewise, how deep ecology is underpinned by a critique of capitalism (Mikulak, 2007).

It is not entirely accurate, therefore, to consider this body of knowledge as a collection of distinct, 'monolithic' causes. Mikulak's combination, for example, would suggest that environmental damage is caused by a capitalist economy, a social class system *and* a human/Nature distinction. This observation helps to re-qualify rather than dismiss Halsey's broader point, which is that these ecological theories all believe that environmental damage can be attributed to one or several universal causes. And, by extension, they believe that tackling these causes will resolve the problem of environmental damage as a whole (Halsey, 2006). This argument has a link with the problems I identified with formal assessments like the Code for Sustainable Homes and BREEAM. The regulation directing such assessments also assumes that unsustainable designs can be identified, that they result from identifiable causes located in the design process and that these causes can be tackled by applying an appropriate combination of assessment and regulation.

The idea of a universal cause and corresponding solution leads me on to Halsey's second point. Halsey notes that, by fixing their perspective around one or several causes, these five ecological theories are unable to 'explain the mobility and meaning(s) of environmental issues' (Halsey, 2006: 37). Halsey illustrates this point by drawing on the environmental issue surrounding the wood-chipping industry in Australia.

Halsey notes that 85% of wood-chippings are sourced from Australia's forests. This activity is supported by government policy but heavily contested by public opinion. Each of the five ecological theories, he argues, fails to appreciate the complexity that surrounds wood-chipping practice, policy, public opinion and their relationship with the material reality of the forest.

Liberal ecologists, he notes, might frame this relationship as a distinction between an expert-led proposal for meeting economic demand in a sustainable way and the views of an uninformed public. However, as Halsey notes, many 'experts' admit that this industry is unsustainable, and economic reports have shown falling demand for these products. An ecomarxist might frame this as a way of putting jobs and profit before environmental sustainability. Yet evidence suggests that these industries have revealed continued economic and employment losses. Halsey illustrates similar contradictions for ecofeminism, social ecology and deep ecology (Halsey, 2006: 36–37).

For Halsey, these ways of falsely explaining environmental damage are not the only area of concern. Similar problems can be seen in the way judgements are formed about environmental damage.

Halsey's first concern relates to the distinction made between human crimes and a non-human environment. At present, most environmental crimes are defined as damage perpetrated by 'individuals and corporations' (Halsey, 2006: 43). But, as Halsey notes, individuals and corporations are not the only, or indeed, the greatest contributors to environmental harm. The state has a much greater impact but is judged in a very different way because it is classified as an institutional body rather than a human body (Halsey, 2006: 44). Like Bonta, Halsey's argument reflects a challenge to the distinction between human and the non-human bodies.[4]

However, for Halsey, the benefits offered by current realist criminology literature are undermined by two ideas. The first results from their connections to ecological theory and the idea of a universal cause and solution. The second concerns the idea that judgement should be made in reference to one or several ideals.

To illustrate this second idea, Halsey reviews three key studies in 'biocentrism' and 'ecocentrism'. In each of their proposals, judgements are made in reference to an essentially defined ideal. For Barnett, environmental crime should be judged against a timeless understanding of 'beauty and integrity' (Barnett, 1999; Halsey, 2006: 46–47). For Benton, judgements should be made against a universal definition for 'human and non-human rights' (Benton, 1998; Halsey, 2006: 49), whilst Lynch and Stretsky refer to a universal understanding of 'environmental justice' (Lynch and Stretsky, 2003; Halsey, 2006: 53).

Halsey's argument against the idea of a universal cause and a universal reference for making judgements is that they prevent us from developing a nuanced understanding and engagement in human/environmental interactions (2006: 53). This argument is important to this study because it reflects the point I raised in Chapters 1 and 2 concerning the use of formal assessments as universal references for making judgements about sustainable design.

Given Halsey's critique of current literature from ecology and criminology, we might ask:

- how else should we conceptualise the effects one assemblage (discussed below) has on another (environmental damage) without referring to overarching, causal explanations?

- how do we determine and judge these effects without a fixed ideal such as 'beauty and integrity', 'human nature' or 'environmental justice'?

Halsey answers these questions by developing two interconnected concepts from Deleuze's philosophy: the *assemblage* and the *machine.*

The core concepts drawn from Deleuze's philosophy

Assemblages, machines and violence

Halsey considers the Goolengook forest as a collection of assemblages operating in what Bonta had termed 'a complex space'. The difference between Halsey's study and Bonta's study becomes visible when one considers how they each discuss the creation and re-creation of assemblages.

Above I showed that Bonta offers us two instances in which an assemblage is created. These can be understood according to the potentials available to them. In the first, an assemblage adds or omits an entity to meet the operational demands of other entities *within* the assemblage (its *multiplicity*). Whilst in the second, an assemblage adds or omits an entity to meet the demands set out by an entity in a different assemblage (a *referential model*). This, I noted, reflects the distinction discussed in Chapters 1 and 2 between assessments within the design process (drawn from *multiplicities*) and formal, essentialist assessments (*referential models*).

Halsey's work, however, seems to suggest that assemblages are all created in a similar way:

> the formation of bodies [or assemblages] (a mineral body, a forest body, a recreational body) cannot be divorced from the enunciations (or process of naming) that brings them into being.
>
> (Halsey, 2006: 97)

This quote illustrates Halsey's belief that the 'process of naming' sets out the conditions through which assemblages are created. Similar arguments can be seen in the work of other spatial theorists (see Robbins and Marks, 2010, for example). Throughout his study, Halsey describes these conditions as a kind of 'violence' because they determine how an assemblage selects what will and will not be made 'visible' or operational (Halsey, 2006). This position is similar to Bonta's description of buffer-zone concepts. Like Bonta, Halsey argues that some concepts introduced into 'complex space' change the way other assemblages are created and re-created.

However, in his review of Halsey's study, Bonta questions Halsey's argument that assemblages are created through the violence of naming machines because it 'gives the false impression that, in the Deleuzian world, nothing is beyond or prior to the text' (Bonta, 2008: 576). This stance taken by Bonta is supported in Deleuze's work. Deleuze's study of geology with Felix Guattari, for example, presents a complex space formed from assemblages created in the absence of human language (Deleuze and Guattari, 2004b: 44–82).

It seems, therefore, that Halsey's suggestion that assemblages are created through the 'processes of naming' should be re-qualified. Instead, we might argue that the process of naming and the 'violence' it causes helps us understand the role humans play in forming a complex space. Rather than showing us the complex interaction between two different ways of creating an assemblage as Bonta's study did, Halsey's study puts these human engagements with complex spaces into critical relief. It is in this sense that I would like to review his work.

Halsey uses these two concepts, assemblage and machine, to explain how and why this environment changes over time. This can be seen in an example from his longitudinal study.

An important change in Halsey's study area begins with a letter written by the Division of Forest Management in 1972. This letter declares the subdivision of the Orbost Forest District into a series of 'Forest management blocks'. For Halsey, this letter initiates a number of 'naming machines' such as 'forest block 21' and 'forest block 22'.

By dividing the land into numbered blocks, Halsey demonstrates how these newly created machines ignored or removed topological features, the presence and interaction of different biological assemblages and the 'multiple histories' that have helped mould the land in different ways (2006: 159). Halsey also describes the way these blocks created the conditions for a new structure to emerge based on the activities and demands of the logging industry. By doing so, the land within any given forest block could be understood in reference to the operational demands of one, dominant assemblage. Not only did these forest block machines limit which assemblages could operate in these spaces; they also created and affected other assemblages operating at different scales and in different environments. Looking across the history of the Goolengook, one can identify a number of such examples.

In 1986, the forest block machine, conceived at the scale of the district, was reproduced at a much larger scale. The Victorian Timber Industry Strategy divided the State of Victoria into fifteen Forest Management Areas (FMA), each containing a number of Forest Management Blocks. These FMAs are discussed according to their ability to supply annual volumes of graded saw logs (2006: 172). Four years later, the 1990 Forests (Timber Harvesting) Act extended this further by charting the total, sustainable volume of logs that could be supplied over a fifteen-year period. It also introduced a requirement for the relevant Minister to undertake a five-year review of sustainable yield calculations (2006: 181–182).

This example reveals a number of machines operating within a complex space. It shows how these machines led to the creation of other complementary or conflicting machines and how they each affect the way different assemblages develop over time. For the most part, this analytical aspect of Halsey's study is successful. The problems with this framework arise when Halsey moves from analysis to engagement: an important move in the context of this book given that both analysis and engagement are integral to planning.

Tolerable and intolerable violence

Halsey's proposals for engaging in complex spaces centre on the way one judges these 'naming machines'. He believes that we should not think of the violence caused by these machines as inherently good or bad (2006: 92). Rather, he argues that we should judge machines by asking, 'what becomings are either facilitated or cast aside in such scenarios?' (2006: 63). Machines that allow an assemblage to explore and develop a wide range of potentials would thus be judged positively, whilst machines that significantly limit these potentials would be judged negatively. This alternative way of judging environmental harm, or 'violence', avoids the use of essential ideals such as 'integrity and beauty', 'human nature' or 'environmental justice'.

To form these judgements, Halsey argues we must 'decide upon the thresholds at which a certain type of [lexical] violence will be permitted [or tolerated], rather than upon how to eradicate the violence of naming *per se*' (2006: 234). For Halsey, tolerable violence occurs when a machine creates conditions that allow assemblages to change in their own way and to pursue new relations with other entities and assemblages (active). Intolerable violence occurs when a machine creates a set of conditions that determine how assemblages will develop (reactive) (2006: 247). It is in this distinction between tolerable and intolerable violence that we can start to see the distinction discussed by Bonta and myself concerning the creation of an assemblage according to a multiplicity or according to a referential model.

To illustrate how this distinction might form a basis for judgement, Halsey returns to the machine outlined above: the naming of the forest into forest blocks. For Halsey, this machine illustrates an instance of extreme or 'monumental' violence because it 'strictly limits the velocities (directions) [the Goolengook assemblage] can chart' (2006: 234).

However, whilst this example is useful, it represents an extreme case. Halsey does not discuss other less significant acts of violence. Yet, such cases are necessary if one is to understand where and how to locate the threshold between tolerable and intolerable violence in this or any other complex space.

The resulting theoretical framework and its gaps

Whilst Halsey's concept of violence and his proposals for assessing violence are useful additions to the concepts drawn from Deleuze's seminal texts, they also lead me onto what I believe to be the principle gap in Halsey's theoretical framework. As noted above, Halsey argues that a naming machine cannot be judged as inherently good or bad but must be judged according to the becomings (potentials) it encourages or blocks. However, this principle is not developed further, and we are left with two other problems.

The first of these concerns the identification of becomings. After all, Halsey suggests that we form judgements based on becomings (potentials) that have been actualised, those that still might be actualised, and those that were blocked or removed. Whilst we might be able to trace the former in the assemblages we see around us (in the actual), the same cannot be said for the others. Halsey offers little indication as to how we should identify these 'pre-actual' becomings.

The second problem concerns assessment. Once identified, Halsey provides us with little indication as to whether we should assess becomings on quantitative or qualitative grounds. If we pursue the former, then a machine would be judged as positive if it encourages more becomings than it blocks. However, this may lead us to questionable conclusions. One can imagine a scenario in which a machine that encourages many racist, sexist and socially repressive becomings is judged in the same way as a machine that encourage many sustainable, egalitarian and socially cohesive becomings.

If one were to resolve this dilemma by pursing a qualitative assessment, then it would suggest judging a machine according to the desirability of the becomings it encourages or blocks. However, the problem here is that we must then decide what is or is not deemed to be desirable, thus returning us to the problem of judgement.

I believe that this problem is owing to the way Halsey (re)creates Deleuze's concepts to form his own theoretical framework. To illustrate my argument, I would like to start by capturing Halsey's framework in tabular form below (Table 3.2).

Table 3.2 Mark Halsey's theoretical framework (Halsey, 2006)

Deleuzian concepts	*Halsey's concepts*
Assemblage: A combination of human and non-human entities.	
Machine: A machine is a collection of forces acting on a number of assemblages. This leads them to structure and re-structure themselves in very particular ways.	**Naming machine:** A naming machine determines how assemblages operate, whether it will develop into the future and whether other assemblages will be created. **Violence:** Violence is the effect that naming machines have on assemblages.
Becomings: Becomings are the potentials for an assemblage to develop in different ways.	**In/tolerable violence:** Tolerable violence occurs when a machine allows the assemblage to pursue a wide range of becomings from its multiplicity. Intolerable violence occurs when a machine significantly reduces becomings to those set out by a referential model.
Becoming (blocks): The Goolengook assemblage can be understood according to a series of developments broken into blocks of becoming.	

The left-hand-side column lists four of Deleuze's key ontological concepts and the way Halsey interprets these concepts to make them useful to his study. Running from top to bottom of the column, these concepts are positioned across the ontological realms that form Deleuze's understanding of the real: the assemblages and the machine as part of the actual, and becomings/blocks of becomings as part of the virtual (or, more precisely, part of the processes of actualisation) (DeLanda, 2002).

The second column to the right-hand side identifies the concepts that Halsey creates to meet the specific demands of his project. As the table suggests, Halsey develops the concept of the *machine* to form the concept of *naming machines* and uses this as the basis for developing the concepts of *violence* and *in/tolerable violence.*

Deleuze's concept of *becoming* is an important part of Halsey's approach to judgement. However, as the table shows, Halsey does not (re)create this concept to meet the specific demands of his field. For me, this provides us with some explanation as to why Halsey's study is unable to identify and measure becomings and thus establish a practical basis for judgement.

The problem is that Halsey's proposals are limited to the way Deleuze uses this concept in his seminal texts. Such texts offer very few clues as to how we should identify or measure *becomings*. One of the most developed examples can be seen in Deleuze's review of Freud's diagnosis for Little Hans (Deleuze and Guattari, 2004a; Deleuze, 2007). Deleuze argues that Freud's psychological judgements focus too heavily on a pre-conceived set of ideas and explanations. In doing so, his analysis of Little Hans does not account for his becomings: becoming-animal, becoming horse (Deleuze, 2007: 98). Neither does he allow for the assemblages that Hans has created from these becomings (Deleuze, 2007: 98). However, Deleuze does not expand on this in detail, leaving us with a partial image of Hans' potentials 'to become' and no way to assess these potentials. Given that Deleuze intends this as a philosophical critique, it is not surprising that he does not develop this concept to meet these demands. However, the same cannot be said of Halsey's study of the Goolengook.

If Halsey is to offer a new, practical tool that can be used to engage in complex spaces, I believe that this gap must be resolved because, to be useful to criminology or to planning, there is a need to move beyond critical analysis and towards engagement.

Rather than using Deleuze's concept of becoming as it appears in his seminal texts, I believe we must (re)create this concept in a way that allows practitioners to identify and assess the becomings that are encouraged or blocked by naming machines. Doing so would make Halsey's concepts of 'tolerable and intolerable violence' of practical use to assessment practice. Given that this is the core focus of my study, such concepts will be taken forward in my subsequent proposals developed in Chapter 6.

Hillier: a multiplanar theory for engagement

I started my review of Bonta and Halsey's studies by outlining a number of empirical observations taken from very specific places. I showed how these empirical observations introduced problems that both theorists believed would benefit from a Deleuzian insight. In both cases we are left with an unstated question: how might this study of a specific case change the way we engage in a range of spatial problems and 'complex spaces'?

This question is fundamental to Jean Hillier's work as a planning theorist. In the introduction to her 2007 monograph, Hillier suggests that her study aims 'to take Deleuzoguattarian geophilosophy beyond the abstract to a useful, practical basis for spatial management' (2007: 13). Hillier's study, therefore, captures the macro-objectives of this study – i.e. to consider whether Deleuze's ontological concepts can be translated into planning theory and planning practice.

This problem at the heart of Hillier's study is much broader and more abstract in focus than the problems underpinning the last two studies. To understand and resolve these problems, Hillier draws on the most abstract parts of Deleuze's philosophy. The use of abstract, philosophical concepts to resolve abstract, spatial problems can make Hillier's work difficult to follow. In order to make this Deleuze-inspired voice accessible and to show its relevance to this book, I will try to unravel some of Hillier's thinking.

A brief introduction to Hillier's study

Like Bonta and Halsey, Hillier forms her proposals from a detailed review of Deleuze's ontological concepts (2007). Such interpretations have attracted the attention of the broader Deleuzian community. These interpretations have also had notable influence in planning theory. Hillier's Deleuze-inspired approach has been used to discuss broader shifts in planning. For Porter and Davoudi, Hillier's proposals are part of a small academic community looking outside traditional planning theory to 'develop more relational, fluid, and interpretive approaches to planning' (2012: 331). Whilst for Searle and Bunker, Hillier's multiplanar theory is an important contribution to a developing 'Australian paradigm'. This paradigm is thought to bring together new theories and practices of strategic planning in metropolitan cities such as Sydney and Melbourne (2010). These approaches, they argue, share the idea that strategic plans must be created to allow for, but not predict, future changes and adaptations.

Hillier's unique contribution to the idea of adaptability has been discussed by a number of other planning theorists. For many, Hillier's study is seen as an experimental and performance-centric approach to planning (Healey, 2009b; Ploger, 2010a; 2010b). For Wilkinson, this idea of experimentation is best captured in Hillier's interpretation and use of the Deleuzian concept 'becomings'. By thinking in terms of becomings, she notes, planners are able to 'navigate towards desired trajectories' and consider endless adaptations to a changing and complex world (2012: 162). This, she goes on, might provide planners with a framework for developing resilient forms of planning practice. Wilkinson's comments in particular show how Hillier's Deleuze-inspired study has since been used to develop other lines of enquiry in response to current debates in planning.

Since Hillier published her proposals for a 'multiplanar' theory in 2007, several attempts have been made to develop and 'test' these ideas in strategic planning practice. Nyseth et al.'s study of Tromsø in Norway provides us with the most developed attempt to date (2009). In this study, the research team compared the approaches to strategic planning in practice with those offered in Hillier's proposals. Their findings suggest that some of the Deleuze-inspired concepts used by Hillier, particularly those associated with becomings, could be identified in existing planning methods.

The problem directing Hillier's study

Hillier's proposals for a multiplanar theory are constructed around a problem she identifies in many areas of planning theory and practice. This problem concerns the use of transcendental principles and models rather than methods and theories based on an immanent and changing world. Whilst Hillier does not offer a detailed review of assessment methods used in the milieu of planning and design, one can see how closely this argument relates to my own line of enquiry presented in Chapters 1 and 2.

Hillier identifies two areas in planning theory and practice where efforts to embrace immanent practices and theories of planning are overshadowed by subsequent attempts to relate these proposals to a transcendental theory. To appreciate this argument, I would like to outline and discuss these two areas in turn.

Consensus building

The first area of planning discussed by Hillier is captured by Judith Innes and David Booher's study of 'consensus building' and the references they make to network science (Innes and Booher, 1999; 1999a; 2010; Booher and Innes, 2002). As the name suggests, 'building consensus' includes both a process (building) and a product of this process (consensus). Innes and Booher explain the process of building consensus through the concept 'bricolage'. As in many disciplines across the social sciences,[5] bricolage is considered as a process of assembling heterogeneous components whose origins lie outside of the specific issue under consideration. In this instance, such components are the fragments of concepts, ideas and strategies used or seen in other projects (see Innes and Booher, 1999a).

Innes and Booher's observations from consensus-building practice suggest that the direction this bricolage process takes and the products of this process (consensus) are constructed *during* the creative act itself (1999a; 2010). There are no pre-defined models or archetypes that decide how this should be done or what the outcomes should be (1999; 1999a). In fact, they insist that effective consensus building can occur only when a group of stakeholders are able to devise and adapt their own rules for specific parts of the process and decide what the final product should look like (1999). The way they describe the immanent process of bricolage, therefore, seems to support the Deleuze-inspired arguments made by Bonta and Halsey as well as the points drawn from my own experiences of design practice in Chapter 1.

However, unlike Bonta and Halsey, Innes and Booher assume that this immanent process of bricolage will lead the group to a mostly harmonious end-state, consensus. For Hillier, this is problematic because there is little reason for us to assume that consensus would develop out of difference (2003). Different viewpoints, she notes, contain conflicts, antagonisms and tensions as much as areas of agreement. For Hillier, this shows how the immanent process of bricolage is stifled by an overarching trust in the transcendental concept, consensus.

One can see parallels between Hillier's argument and Bonta's observations. If we think back to Bonta's study of the Olancho region, there is no reason to believe that the interactions between different assemblages like the coffee farm and the cattle farm would lead to a harmonious end-state. Indeed, for Bonta, the complex space that results from these interactions is formed out of disharmony.[6] Similarly, there is no reason to assume that the sequence of design experiments discussed in Chapter 1 should fit together to form a perfectly functional building (hence the resulting problems associated with furniture layouts, corridor design and so forth).

For Hillier, another important area of contention concerns Innes and Booher's attempts to explain and support this idea of consensus. Innes and Booher draw on advances in network theory and structuration theory to explain why consensus is achieved from the immanent process of bricolage.

The use of network theory as an explanatory framework can be seen in most of Innes and Booher's work. Not only are network dynamics used to explain how consensus is achieved between different viewpoints (1999a); it is also used as a model for new ways to collaborate in planning and for institutional change more broadly (Booher and Innes, 2002; Innes and Booher, 2010).

To add further support to this overarching system of order, they argue that the distribution and production of power in the network model corresponds to broader theories of structural dynamics. In this case, they relate power in a network to Giddens' three typologies of power: the power of action; the power of ideas, modes and methods; and the power of deep structures (Giddens, 1984; Booher and Innes, 2002: 225).

For Hillier, Innes and Booher's work captures an inconsistent position. They reject models and archetypes in favour of bricolage: an immanent process of creation (drawn from multiplicities). Yet they introduce pre-defined end points within this process and justify the results using explanatory models and universal theories (referential models).

Relational complexity

The second area of planning identified by Hillier brings together Patsy Healey's arguments for relational complexity and new-institutionalist theory (1999; 2003; 2004a; 2004b; 2006; 2007).

Healey argues that planners must embrace the idea that they too are embedded in the fabric of society (1999). As such, they must see planning practice as a way to work *within* rather than outside of a complex world. Such practices, she goes on, must be undertaken as experiments in an uncertain, dynamic and fluid set of relations (2003; 2004b; 2007).

Like Innes and Booher's idea of bricolage, Healey's proposals for experimentation seem to suggest that the process and products of planning practice should be created immanently – i.e. from *within* the complexity that surrounds us. Yet, like Innes and Booher, this engagement with complexity is set against a theoretical framework developed around networks and structuration theory.

For Healey, geography is 'relational'. Healey describes this geography as a complex collection of overlapping, 'loosely-coupled webs' formed from 'nodes, links and loose threads' (2007: 222). Whilst Healey's description seems to embrace greater degrees of complexity, it is based on the same assumption that the network model can be used to understand the actual world around us. Like Innes and Booher, Healey also assumes that this referential model of the network can be transposed between different uses. Indeed, this is critical to Healey's position.

This has ontological implications. It suggests that all entities and the relationships between entities share a common structure. We might argue that such an idea lends itself to an essentialist stance, positioned at the level of the actual world as a whole rather than at the level of the entities themselves.

For Hillier, Healey's relational space introduces immanent processes of experimentation (drawing on multiplicities). But, like Innes and Booher, she goes on to explain this experimentation by referring to network models and universal theories (reference models).

Hillier uses these two examples from planning theory to highlight a broader problem in the discipline. In philosophical terms, she questions why many planning theorists and practitioners advocate *immanent* practices but explain these by referring to *transcendental* models and theories.

For Hillier, this problem cannot be resolved by simply discarding transcendental structures. After all, these structures do more than simply explain the way plans are made. They have functional roles in planning practice. Planning professionals would be unable to form or enforce their plans without a set of statutory regulations for example.

The solution, she argues, is to distinguish between these two kinds of transcendental structure: those that try to explain complexity, and those that function within complexity. For Hillier, spatial planning theory and practice needs a theoretical framework that takes account of solutions that are created immanently. Such a framework, she argues, must avoid any reference to simple models and structural dynamics. For Hillier, it is in the construction of such a framework that Deleuze's philosophy can be of greatest benefit to spatial planners.

The core concepts drawn from Deleuze's philosophy

As in Bonta and Halsey's studies, Hillier's proposals for planning are based on the way she interprets, expands and (re)creates some of Deleuze's key ontological concepts. Whilst most of the concepts in Bonta and Halsey's studies focus on the actual, Hillier develops her multiplanar theory around a number of concepts within Deleuze's virtual, or 'pre-actual', realm (DeLanda, 2002). This is not entirely unsurprising given that the role of planning is to create plans based on 'what might become', rather than what has already become.

Plan(e)s

Hillier develops her theory around Deleuze's concept of *the plane* and its usefulness to spatial planning. To do this, Hillier looks to the explanations offered by the Deleuzian scholar Brian Massumi. In his translation notes for *A Thousand Plateaus*, Massumi explains that the French term '*le plan* designates both a "plane" in the geometrical sense and a "plan" ' (2004: xvii–xviii). Whilst Deleuze and Guattari primarily use *le plan* to mean 'plane', there are times when both meanings are intended, such as during their discussion on the plane of transcendence.[7] In such instances, Massumi makes this distinction by using the term 'plan(e)'.

Rather than using the dual meaning for only one kind of plane, Hillier extends this to include both planes. Thus, for Hillier the plan(e) of immanence and the plan(e) of transcendence can be seen as geometric planes in a theoretical framework, and as plans more specifically (2007: 242).

Plan(e) of immanence: a figure of the plane/strategic plan

By drawing on this dual meaning, Hillier begins to forge links between philosophical planes and spatial plans. To do this, she must interpret and adapt some of Deleuze's explanations.

Hillier introduces Deleuze's concept of the plane of immanence as a 'virtual realm of potentials' (2007: 244). Because it is virtual, Hillier argues, it will always remain 'inaccessible to actants such as spatial planners' (2007: 246). However, this should 'not stop us "figuring it, or constructing images of it" ' (2007: 246). Rather than focusing on the virtual plane, Hillier focuses her attention on this 'figure'. In doing so, she reflects the point I made in Chapter 2 concerning Deleuze's use of extremes like 'the virtual realm' to encourage theorists to think about processes of actualisation.

This position is equally supported in Bonta and Protevi's proposals for a 'geophilosophy' (2004). Like Hillier, Bonta and Protevi form a distinction between two planes of immanence. They describe the first as a '(relative) plane' and the second as 'THE virtual plane' (2004). Like Hillier, they distinguish these according to our ability to construct and understand them:

> in general humans have the capacity to transform and re-smooth their landscapes by transmuting their spaces, drawing new 'planes of consistency' [immanence] for them.
>
> (Bonta and Protevi, 2004: 173)

Whilst they are set in very different contexts and use different terms, this comment seems to offer some support to the links Hillier draws between a figure of the plane of immanence and a strategic plan (Hillier, 2007: 249).

Becomings and speculative potentials

Hillier develops this figure of the plane of immanence (strategic plan) by focusing her attention on the becomings that form it. A plan(e) of immanence, she notes, is formed as a 'transvaluative, collective speculation about what might be' (2007: 243). In this, Hillier confirms Deleuze and Guattari's point that the plane of immanence is made up of potentials. In other words, it is formed from potentials *to become* something/s rather than a collection of things that have already been actualised *into* something (entities).

This description is made all the more interesting by the two terms she adds to this definition: 'collective speculation' and the 'transvaluative'. For Hillier, a figure of the plane is formed from a set of potentials imagined by a group of 'actants' working together. These actants may include local residents, planning consultants, masterplanners, architects, builders and policymakers.

Hillier's phrase also suggests *how* these people might construct this figure. By referring to 'transvaluative' potentials, Hillier distinguishes this figuring from other forms of brainstorming or collaborative work. This term implies that the group of actants must imagine potentials without relating them to existing or assumed principles. Instead, they should make their judgements according to 'the forces that intersect it and the things it can do' (Kaufman, 1998; Hillier, 2007: 243; Hillier, 2011). Building on Kaufman's description, Hillier describes these forces as the relationships between potentials and the way these potentials might transform established relationships (2011: 508).

Thinking back to Hillier's critique of consensus building, one can see why this is important. In Innes and Booher's practice of consensus building, a facilitator encourages a group of stakeholders to consider potentials that might help them to achieve consensus (1999; 1999a). In such instances, consensus is taken as an end-state, and potentials are valued according to the way they might contribute to achieving this.

In Hillier's proposal, potentials are considered according to their capacity to affect the relationship between entities. In doing so, she broadens the likely directions such collaboration might take.

Empty signifiers

However, this raises an important point concerning *how* one should speculate, or *what* one should speculate about. In her final description for the figure of the plane of immanence, Hillier discusses

> [s]everal (or perhaps one collectively preferred) trajectories or 'visions' of the long-term future, including concepts towards which actants desire to move such as sustainability (plan(e)s of consistency or immanence).
>
> (Hillier, 2007: 249)

For Hillier, these empty signifier concepts provide a focal point for 'speculating becomings'. Thus, actants would observe the entities that surround them and consider what potential these entities have to become 'sustainable', or 'more socially cohesive', and so forth.

Hillier's proposal to include concepts like 'sustainability' into her theory is drawn from her work with Michael Gunder and their Lacanian/Zizekian analysis of spatial planning (2004; 2007; 2009). In one such publication they identify ten concepts seen across planning theories and practices that have

> given up explicit, concise, significance to secure multifarious points of view, chains of significations constituting conflicting narratives, or unique interpretations pertaining to particular situations, all under one common label.
>
> (Gunder and Hillier, 2009: 17)

Such open-ended concepts are described using the Lacanian term 'master [or empty] signifier'. As in Lacan's analysis of the individual, Gunder and Hillier argue that planning theories and practices are formed from a combination of terms used to construct an identity. Thus, a spatial policy document, for example, may draw on concepts like 'sustainability', 'certainty', 'growth' and 'globalisation' to help identify what the policy is and how it relates to other policies and debates in the field. Yet the meaning of these concepts and the relationships between concepts are never explicit. They are always open to different interpretations, or points of view. Drawing on Laclau, Gunder and Hillier refer to these concepts as 'empty' or 'floating' concepts (Laclau, 1996; 2003; 2005).

Whilst Hillier does not make the connection, this role played by 'empty signifiers' is not unlike the role played by singularities within Deleuze's seminal texts and outlined in Chapter 2 (DeLanda, 2002). Like empty signifiers, singularities provide a direction for becomings as they develop from the virtual towards the actual, or, in Hillier's framework, as they move from the figure of the plane of immanence towards the actual world we see around us (DeLanda, 2002: 14). This idea is developed further in Chapter 6.

Emergence: the un-speculated potentials

Deleuze describes the plane of immanence as a 'transcendental field' (Deleuze, 2001: 25): a virtual realm inaccessible to our sensory observations (hence 'transcendental empiricism' Bryant, 2008a).[8] Thus, whilst we may be able to overcome some of these limitations by speculating potentials for the future, Hillier reminds us that the resulting figure is inevitably partial:

> [A plane of immanence] is a plane of foresight; of trajectory, of creative transformation, of what *might* be. Chance is important, however. We should not forget the potential for unforeseen lines of flight to emerge.
>
> (Hillier, 2007: 245)

Hillier develops this idea of emergence through un-speculated potentials when she notes that 'there is thus much scope for things to not turn out as planned; for something to emerge in the gap between plan and built form, between virtual and actual' (Hillier and Abrahams, 2014: 33). In doing so, Hillier warns us that we must see strategic plans as temporal – open to revision both at different periods in time and at different scales of plan-making (Hillier, 2011).

Plan(e) of transcendence

Hillier does not offer speculated and un-speculated becomings as the only factors influencing the actualisation of assemblages. As a counterpoint to the plan(e) of immanence, Hillier suggests that these assemblages are also created in reference to the plan(e) of transcendence. As with the plan(e) of immanence, Hillier describes the plan(e) of transcendence both as a geometrical plane in an ontological framework and as a plan from the field of spatial planning. If a strategic plan is an example of the former, she argues, 'local plans, design briefs [and] detailed plans are typical planes of [transcendence]' (Hillier, 2007: 247). Unlike strategic plans, these plans relate specifically to the way we organise, define and construct space.

For Hillier, such plans act as 'masterplans' or 'blueprints . . . with certain goals for development', and they set the standards through which subsequent decisions are judged (Hillier, 2007: 247; Hillier and Abrahams, 2013). Thus, a local plan will set goals for how many homes will be constructed in a given part of a town. It will also set the standards on how these houses will be built in terms of heights, use of materials, relationships to context, number of car parking spaces per house and so on. And it will use these to influence the design and location of specific plans, such as a masterplan or a building plan.

Thinking back to Chapters 1 and 2, we might argue that my description of the essentialist, modelled concept is similar to Hillier's description of a transcendental plan(e). It sets out a series of goals based on carbon neutrality, it defines 'blueprints' for achieving this through a universal assessment method and it sets standards by which subsequent decisions will be judged. Hillier's study suggests that these modelled concepts are an inevitable part of the regulatory and planning/design process. She does not suggest that these modelled concepts should be removed in all instances, because her study infers that they serve a useful purpose.[9] However, for Hillier, it seems that modelled concepts become problematic when they are used as a fixed goal against which all subsequent decisions are judged or organised.

The resulting theoretical framework and its gaps

As with Bonta's and Halsey's theoretical framework I have captured Hillier's theoretical framework in tabular form (Table 3.3). The left-hand-side column identifies the ontological concepts drawn from and interpreted by Hillier in her proposed framework. Down the column, I have shown how many of Hillier's concepts are created within the virtual or 'pre-actual' area of Deleuze's ontology, the extent of which reflects Hillier's underlying message to 'stretch beyond the

Table 3.3 Jean Hillier's theoretical framework (Hillier, 2007)

Deleuzian concepts	*Hillier's concepts*
Assemblage: An assemblage is an actualised group of potentials.	
Plane of transcendence: A plane of transcendence is a collection of blueprints used to explain complex relations.	**Plan(e) of transcendence:** The plan(e) of transcendence is a blueprint for a given area or an area of design.
Becomings: A becoming is the potential for an entity to develop in the future.	**Un/speculated potentials:** Speculated potentials are potentials that can be imagined. Un-speculated potentials are potentials that cannot be imagined.
Becoming (blocks): Blocks of becoming are a collection of potentials directed towards the development of an assemblage/s.	**Empty signifiers:** Empty signifiers have no inherent meaning. They provide direction to our speculations about what 'might be'.
Plane of immanence: A plane of immanence is a chain of multiplicities (virtual diagrams).	**Plan(e) of immanence:** The plan(e) of immanence is the combination of all speculated potentials directed towards a series of empty signifiers already given within the field.
	The strategic plan: A strategic plan is an example of a figure of the plane of immanence.

horizon' or, in other words, stretch beyond the actual world we see around us. We can also see that, whilst they are used as part of a very different project, these ontological concepts are used and described in similar ways to those set out by Bonta, Halsey and myself in Chapter 2.

The right-hand-side column shows concepts created by Hillier. As in Bonta and Halsey's theoretical frameworks, this table demonstrates how Hillier interprets, expands and (re)creates Deleuze's concepts to meet the unique demands of her field.

This table also highlights what Hillier suggests as three influences leading to the creation of a planned or designed assemblage:

1 the first group is offered the greatest attention and centres around the strategic plan (plan[e] of immanence), formed from speculated potentials directed towards empty signifiers (or singularities);
2 the second group is the un-speculated potentials that emerge over time and as the design process moves across different kinds of plan and stages in the

actualisation of a city or building, from strategic plans to masterplans, building plans and the built environment;

3 the third group is defined by the plan(e) of transcendence that determines how space should be structured (striated) through blueprints for a specific geographical location or for a specific area of design (such as established layouts for WCs, houses or cul-de-sacs).

Thus, Hillier suggests that, as we move from the strategic plan to the masterplan, the building plan, the window jamb detail, a sketch in a wet site hut and the built assemblages we see in the world around us, we must work with these three groups of influences: speculated potentials, un-speculated potentials and blueprints. However, this leaves us with a series of difficult and mostly unresolved questions important to this book:

1 how do the masterplanners or architects sat at their desks and drawing boards take Hillier's strategic plan (formed from speculated potentials) and use it as a practical basis for their proposed designs: to direct the processes of actualisation?
2 how should they combine these speculated potentials with the un-speculated potentials that arise during the process of plan-making (such as the potential for poor ground stability identified in the example used to introduce this book)?
3 how should they negotiate these influences with the influence of regulations and blueprints and, what I term, 'modelled concept-based tools'?

Hillier's multiplanar theory is mostly focused on the creation of a strategic plan, rather than questions of assessment, design and actualisation. Of course, this is not unsurprising given that Hillier's work is situated within the field of strategic planning rather than the field of planning/design. However, if we are to make Deleuze's ontology useful to plan-making practices more broadly, then I believe we must 'stretch' Hillier's multiplanar theory in both directions: 'beyond *and* towards the horizon'. Or, in other words, we must direct our speculations towards the virtual, but we must equally consider how these speculations might be actualised and assessed through the many decisions we make in different areas of plan-making practice. This, I believe, highlights a gap in Hillier's framework.

Hillier has started to offer some initial and tentative responses to these questions. The first of these can be seen in her attempts to identify 'approximations of multiplanar planning in practice' (Hillier, 2007: 275). These examples from practice include reference to the European Spatial Development Perspective (ESDP), municipal planning in Norway and proposals for Melbourne Docklands (2007: 276–291).

Hillier selects these cases in particular because they share a mixture of strategic planning and local planning proposals. Each example includes instances when strategic planners have experimented with new methods and have left parts of their strategy open to subsequent interpretation and development by actors operating at the local level. In her review of the ESDP, Hillier notes that it is a

> plane of immanence upon which more detailed and different existential sphere and place-specific plans and policies emerge, where the voices of local communities may be heard and political power-games played out.
>
> (Hillier, 2007: 280)

This quote suggests that Deleuze's concepts can be identified in existing forms of strategic planning practice. We might argue that such links between Deleuzian theory and practice are important if we are to develop a viable and practicable set of Deleuze-inspired planning tools. However, the problem is that once Hillier has established these links, she does not consider how the ESDP, the Norwegian municipal plan or Australian docklands can usefully contribute to a Deleuzian alternative to planning, or an alternative set of tools used by planners. As a result, Hillier's proposal acts as an analytical frame for existing practice rather than something that might *change* existing planning practice.

Hillier's subsequent work makes some attempt to resolve this problem. In her 2011 paper, Hillier establishes conceptual links between her multiplanar theory and Richard Hames' methods of strategic navigation. Such links are supported by planning theorists using this method in practice (Wilkinson, 2011: 598). Like Hillier, Hames also focuses on potentials for the future, experimentation and adaptability. But these principles are not strictly Deleuzian and can be found in a range of literature in many disciplines. As with her review of the ESDP, Hillier identifies links with existing forms of practice but does not critically discuss this link or consider how it might advance her own proposals for a new set of practicable planning tools.

Hillier's study suggests, therefore, that Deleuze's philosophy can be used to help planners re-think key parts of their practice. Yet her proposals for a new Deleuzian set of planning tools intended for engagement rather than analysis remain incomplete. Whilst Hillier does not develop her concepts fully, they provide us with useful additions to the concepts offered in Deleuze's seminal text. These concepts are particularly important to professionals engaged in the field of planning/design, as well as those engaged in the assessment of building and urban designs. Given that this is the core focus of my study, such concepts will be taken forward in my subsequent proposals developed in Chapter 6.

Discussion

These three studies provide some directions for answering my two research questions and a number of broader lessons about making Deleuze useful. These will be discussed below according to the following three themes:

1. Directions for responding to the essentialist, 'modelled concept'
2. Is Deleuze useful as a method of analysis or/and engagement?
3. Lessons about making Deleuze's concepts useful.

Directions for responding to the essentialist, 'modelled concept'

In this first theme, I explore how these three studies respond to the idea that a concept can be modelled according to its essential traits. Whilst these three Deleuze-inspired studies do not address 'modelled concepts' directly, they do make references to comparable concepts and associated problems. Taking these into consideration, I consider what can be learned from these and what directions they offer for developing a suitable response to the modelled concepts introduced and discussed in Chapters 1 and 2.

Bonta's study illustrates two kinds of entity operating in a 'complex space'. The first is illustrated in his description of roaming cattle and the subsequent introduction of fencing. In my review, I showed that, whilst the cattle affected the coffee farm assemblage, it was not responsible for deciding how the assemblage coped with these affects (i.e. the introduction of the fence). This fence was introduced into an assemblage to meet the operational demands of other entities within that assemblage and to respond to the effects of 'entanglement'.

The second kind of entity is illustrated in his description of 'buffer zone concepts' as promoted by the conservation assemblage. As I noted above, Bonta informs us that this concept was 'based on the US model . . . the "perfect" park out of the US textbook' (Bonta and Protevi, 2004: 183–184). It was, we might argue, a model created within a different context, country, time and purpose. This second kind of entity was aimed at changing the way other assemblages operated. They did this, he notes, by determining how other assemblages organised themselves and the entities they selected or omitted from the assemblage.

This 'buffer zone concept' as described by Bonta is comparable to what I termed 'modelled concepts'. As in my reference to the 'sustainable home' or 'sustainable, newly constructed building', they too capture the idea that a concept (in this case 'the perfect park') can be modelled to form a set of principles transferable to many different spaces. As in modelled concepts, these principles are also used to determine how a space (or design/plan) will evolve in the future.

Interestingly, Bonta's analysis suggests that, once introduced, these modelled concepts become part of a complex space. Bonta's principle concern, it seems, is that these modelled concepts should not be allowed to provide the *only* means through which other entities are introduced into a complex space. The different assemblages that form a complex space, he suggests, should be free to introduce entities that meet their unique operational demands also. As in the case of the roaming cattle and coffee farm assemblage, Bonta seems to suggest that these entities are important to the way assemblages adapt to a changing environment. Thus, rather than removing essentialist, modelled concepts, like those used in BREEAM forms of assessment, Bonta suggests we must prevent them from dominating a complex space (and the design process) and thus limiting the development of other potentials in the future.

Similar links to 'modelled concept' can be identified in Halsey's study of the Goolengook. In his timeline for this complex space, Halsey discusses the introduction and development of the forest block. Halsey shows how this ideal and rather abstract concept was operationalised through a series of other more tangible concepts. As in Bonta's study, these concepts were used to sub-divide land in Goolengook according to pre-defined categories, defining which assemblages would operate in different parts of the forest and which entities would need to be added or omitted from those assemblages.

Like Bonta, Halsey is critical of these concepts and suggests that some form of action is needed to prevent them from dominating complex spaces. Unfortunately, Halsey does not discuss which kinds of actions would be deemed suitable. However, he does argue that any such actions should be underpinned by a process of judgement.

Rather than judging a concept as inherently good or bad, he argues, every concept should be assessed according to whether they encourage or block an assemblage's potentials to develop differently in the future. This argument suggests that, whilst Halsey judges the forest block concept as 'intolerable', other modelled concepts with a lesser effect might be judged more favourably. It also suggests that the 'forest block' could have a positive effect if introduced and developed in different ways (as part of a different machine).

The idea that we should assess modelled concepts on an individual basis is not fully supported by Hillier's study. In my review of her study, I suggested that 'modelled concepts' are captured, at least in part, by Hillier's argument against transcendental concepts/ideas. Such concepts, she notes, arise when a concept like sustainability is defined as a 'desired societal goal' and used to 'coalesce . . . a field of diverse issues' (Gunder and Hillier, 2009: 144). Hillier argues that these transcendental concepts should be removed from the early strategic planning process. Such concepts, she notes, should be replaced with 'empty signifiers' whose meaning is not determined in advance.

Whilst Hillier dismisses such concepts as part of the strategic planning process, she acknowledges that the codes and assessments used in more detailed stages of planning are part of a 'transcendental plane' that cannot be removed from the planning process entirely. And, whilst she does not explain what useful contribution these codes and assessments make, her proposal does infer that they have some tangible benefits to regulatory planning and design.

This leaves us with a mixed proposal for responding to modelled concepts and modelled concept-based tools in Hillier's framework. It suggests that essentialist modelled concepts should not be used as part of early planning (strategic planning) processes but cannot be avoided in subsequent stages of planning (and design). As yet, Hillier has not provided planners, masterplanners and architects with any clear indication as to how they should work with or respond to this latter group of modelled concepts.

All three studies present us with the idea that an alternative to the modelled concept either exists or could exist. These three studies discuss this alternative as something that encourages and reveals an assemblage's potentials (becomings), so that new assemblages or adaptations will emerge from an assemblage's multiplicity in a constantly changing environment.

Whilst these three studies share a common description, they offer different directions for developing an alternative to the modelled concept. These can be summarised as follows:

1 *Balancing*: Bonta suggests that modelled concepts are part of complex spaces. Rather than removing these modelled concepts, and the assessments built from them, we should look for ways to balance them against alternative ways of assessing an assemblage.
2 *A judgement-led response*: Halsey suggests that each modelled concept, and thus each formal assessment, should be assessed on an individual basis. This assessment should be focused on the potentials this assessment encourages or blocks. Any subsequent action should be determined in response to these assessments.
3 *Removing modelled concepts in early plan-making processes*: Hillier suggests that modelled concept should be removed from early strategic stages of planning/design and assessment. They should be replaced with an immanent approach based on speculating potentials for the future.

Is Deleuze useful as a method of analysis or/and engagement?

Unsurprisingly, these three directions relate to the objectives underpinning each study. Bonta's suggestions favour analysis, Halsey's suggestions favour analysis-led engagement and Hillier's suggestions are focused almost exclusively on ways to engage in complex spaces during the early strategic stages of planning and assessment.

Understanding the success or failure of these three objectives is important to the aims of this book. As I noted in the introductory chapter, this study considers whether Deleuze's philosophy can be used to develop a new non-essentialist tool for planning. This alternative is not intended to analyse existing ways of assessing a building or urban plan. Rather, it is intended as a practical tool that can assist planning-related processes in practice.

In the following I bring together the criticisms raised in my review of each study and use these to ask whether Deleuze has been most successful when used to analyse or engage in complex spaces.

My review of Bonta's study identified one principle area of concern. Bonta's concept of 'complex space', I argued, failed to include spaces in countries with highly developed state assemblages. This is drawn from his description of the state as a coherent assemblage. Yet this view has little support in planning and political literature (Degeling, 1995; Cowell and Martin, 2002; Cowell and Owens, 2002; Bevir and Rhodes, 2003; 2006).

My critique of Bonta's 'complex space' concept provides scope for broadening his analytical approach rather than challenging its use in his study of the Honduras. With this in mind, it is fair to conclude that Bonta's use of Deleuze's (geo) philosophy successfully meets his study objectives, which are to help him analyse and understand land use in this complex space.

A similar statement could be applied to the analytical part of Halsey's study. My principle critique of Halsey's analysis of the Goolengook forest is drawn from a comment made in Bonta's review of this study. As I noted above, Bonta argues that Halsey's concept of 'the naming machine' fails to acknowledge the role played by other non-textual machines. As in Bonta's study, this critique provides scope for broadening Halsey's study rather than pointing to a fundamental problem in his analytical approach.

Whilst Halsey's theoretical framework is largely successful in the analysis of environmental damage (violence), it fails in its second objective, which is to offer an effective way to engage in the world through new forms of judgement. Halsey's study offers only a vague principle for his alternative to current essentialist approaches to judgement. As discussed in the last theme, Halsey argues that a machine cannot be judged as inherently good or bad but must be judged according to the becomings (potentials) it encourages or blocks. However, this principle is not developed further, and we are left with two other problems: identifying becomings and assessing these becomings. Thus, it seems that Halsey's attempt to engage in complex spatial problems remains vague and under-worked.

I raise similar concerns over Hillier's proposals for engagement. Hillier's 'multi-planar theory' is largely successful in translating Deleuze's ideas from philosophy to planning theory. Hillier's theoretical framework, for example, relates Deleuze's *plane of immanence* and *plane of transcendence* to different kinds of plan, and the concepts of *becomings* and *blocks of becomings* to the process of plan-making. The latter roughly reflects the outline I presented for the design process in Chapter 2.

However, the limitations of this philosophy/planning transition are made clear in Hillier's subsequent efforts to translate this framework from planning theory to planning practice. She argues that her theoretical framework, and particularly the relationship between the plan(e) of immanence and the plan(e) of transcendence, can already be identified in existing planning practice. However she makes no subsequent effort to develop these links as part of a new Deleuze-inspired set of tools for practice. Rather than using Deleuze's philosophy to engage in planning, this suggests it be used to analyse and justify the way we *already* engage in practice.

These three studies suggest, therefore, that Deleuze's philosophy has been more successful when used to analyse spatial problems than it has to engage in spatial problems. This reaffirms John Forester's concerns that planning theory often draws on philosophical concepts to analyse and critique existing forms of practice but offers very few suggestions for replacing these practices with new tools of engagement (Forester, 2007).[10]

Lessons about making Deleuze's concepts useful

In my review of the three studies, I have shown how important Deleuzian concepts like the *assemblage, becomings/blocks of becoming, and the plane of immanence* have been in the development of each theoretical framework. The tables used to conclude each review show how these concepts are interpreted to meet the unique demands of the field and the problems each theorist identifies in that field. These tables also show how these concepts, and other related concepts, are expanded and (re)created in each study. As my review shows, it is here where most of the unanswered questions and gaps lie.

The problems I identify in Bonta's study centre around his concept of 'complex spaces'. The problems I identify in Halsey's study result from his failure to (re) create the concept of becoming to correspond with his concept of tolerable and intolerable violence. Likewise, Hillier's failure to develop her proposals into a practicable tool is owing to her failure to (re)create these concepts to suit the demands of planning practice.

With this in mind, I suggest that, if Deleuzian theorists are to make the transition from philosophy to planning theory *and* practice, they must be prepared to (re)create Deleuze's concepts many times and in many ways to suit the unique demands of their field (Abrahams, 2013). This argument will be used to direct my own proposals, as discussed in greater detail in Chapter 5.

Notes

1 Bonta uses the phrase 'plugged into' to describe instances where a person or an object engages with, or operates within, a set of relations and ways of doing things.
2 For the purposes of this text, 'becoming' should be taken as 'in transition to'.
3 Examples of such mapping can be seen in Bonta's study explored in more detail below.
4 Indeed, this challenge can be seen in many so-called 'realist' positions including those inspired by Deleuze's philosophy and those inspired by Actor Network Theory.
5 See Lincoln and Denzin (2003) for example.
6 Which Bonta and Protevi term 'far from equilibrium states'.
7 Or 'plane of organisation', which is taken to be approximately synonymous (DeLanda, 2002; Hillier, 2007: 242).
8 See Chapter 7 for a detailed review of transcendental empiricism, how it differs from other epistemological stances and how it influenced my research design.
9 Although she does not offer a detailed review or explanation as to what this purpose is or when and how it is beneficial.
10 This is discussed in Chapter 9.

Part 3

A case study of BRE assessments

Across the last two parts of this book, I have looked to define the outlines of a problem, to review the debates surrounding this problem, to identify directions for resolving the problem and to understand whether Deleuze's concepts can be made useful to engagement as well as analysis. In this third part, I use these lessons and insights to explore the possibilities of making Deleuze useful to a specific area of planning practice. As with most case studies, I do this in three broad sections: selection of a case, methodology and results/analysis. However, unlike more 'traditional' studies in the social sciences, this case study is not weighted in favour of empirical enquiry. Rather, it is formed from two roughly equal and connected lines of enquiry: a theoretical study into a new, Deleuze-inspired assessment tool; and testing the practical viability of the tool through empirical enquiry.

4 A research strategy

Aim and structure of the chapter

The aim of this chapter is to set out the research strategy and research design for this third part of the book. To achieve these aims, this chapter starts by explaining why, of all the modelled concepts in planning, I have decided to stick with the example used to introduce this study. In the second part of the chapter, I outline key stages of the research used to answer the two research questions identified in the introduction – namely:

1 can Deleuze's philosophical concepts be translated into a new theoretical framework for constructing formal assessments?
2 if so, can this be translated into a practical tool useful to actors in the planning and development process?

The third and final part of this chapter draws on Deleuze's epistemological stance and the methods used by other Deleuze-inspired researchers to translate this strategy into two research stages.

Selecting a case study of the modelled concept

As noted in Chapters 1 and 2, essentialist modelled-concept-based tools can be seen in specific planning/design milieu as well as in the broader field of planning. These observations suggest that the case selected for study must offer *specific* conclusions useful to the selected area of assessment, as well as *generalisable* conclusions about essentialism in planning and about the usefulness of philosophical concepts to spatial disciplines.

I used these concerns to formulate two selection criteria:

1 *Criterion 1: a good example of the modelled concept in planning practice.* The assessment tool selected should be one of the most developed and consistent uses of the modelled concept in planning.
2 *Criterion 2: a critical example of the modelled concept in planning practice.* The tool selected should be a critical example of a modelled concept. In the

case of this book, a critical case is taken to be an assessment that significantly affects the way design decisions are made in a specific milieu – an assessment that dominates an area of regulatory practice.

As part of the selection process, a number of assessments were considered and dismissed, the most notable of which are the Strategic Environmental Assessment (SEA) and Territorial Impact Assessment (TIA). The first of these, SEA, was discussed in Chapter 1. As Fischer notes, this assessment is structured around a pre-conceived idea of 'sustainable development', which is used as the basis for systematic evaluation (Fischer, 2003). Fischer's comments suggest that SEAs are designed around a modelled concept (a modelled concept of 'sustainable development') and, thus, broadly meet the first selection criterion. However, Fischer also argues that, since its inception in the 1980s, a range of in-use variations of SEA have developed (Fischer and Seaton, 2002; Fischer and Gazzola, 2006; Fischer, 2010; Da Silva et al., 2014). Whilst some diversity might be considered an inevitable part of the assessment process, Fischer argues that these differences are significant enough to limit the systematic, cross-comparative principles that underpin the assessment's design (Fischer and Seaton, 2002; Fischer and Gazzola, 2006; Fischer, 2010). Fischer's comments point to important inconsistencies in the role played by modelled concepts, thus excluding them as 'good' examples of the modelled concept used in practice (Criterion 2).

My reasons for dismissing TIA as a candidate for the case study lies in its incompatibility to the second criterion. Whilst modelled concepts have been intrinsic to the design of TIA to date (Abrahams, 2013), these assessments have still to be finalised and used across the European Union's member states (ESPON, 2012). It is fair to conclude therefore that, in their present state of development, these assessments do not dominate a specific area of regulatory practice (Criterion 2).

After reviewing a number of such candidates for case study selection, I found that the family of assessments developed by the Building Research Establishment (BRE) provided a strong 'fit' to the selection criteria above. This family includes *the Code for Sustainable Homes* used to introduce this book as well as *BREEAM New Construction* (outlined in Chapter 2), *BREEAM In-Use*, *BREEAM Refurbishment* and *BREEAM Communities*. The following text demonstrates how this group of assessments meet the selection criteria.

Criterion 1: a good example of the modelled concept in planning practice. This broad group of BRE sustainable assessment tools are some of the most developed and consistent uses of the essentialist modelled concept to date. In Chapter 2 I provided a diagram that showed how the modelled concept underpins the Code for Sustainable Homes form of assessment and BREEAM New Construction. A similar schematic can be seen in all assessments in this family.

In Chapter 2 I outlined a number of other assessment tools constructed around essentialist modelled concepts. These included European Territorial Cohesion Indicators (ETCI), the Index of Multiple Deprivation (IMD), Heritage Impact Assessments (HIA) and Housing Quality Indicators (HQI). Whilst these tools are good examples of this phenomenon, none of them have used the essentialist modelled concept as consistently as the BRE suite. This BRE suite of assessments, therefore, acts as a good case of modelled concept-based tools used in planning and, thus, as an ideal opportunity to consider and test an alternative.

Criterion 2: a critical example of the modelled concept in planning practice. The use of critical cases to test a theory or proposal is an established method within case study literature (Yin, 2009: 47–48). Not only are the BRE assessments some of the most developed examples of modelled concept-based tools in planning, but, taken together, they are also some of the most critical examples used in practice.

Whilst these assessments are not mandatory, many local authorities and public sector clients across the UK specify this form of assessment as a condition of planning approval and funding (BRE Global, 2011). According to a survey undertaken by BRE Global, over '40% of local authorities in the UK are specifying sustainable building policies in their plans' (BRE Global, 2011). Drawing from this data, BRE Global argues that such assessments will continue to 'shape the nature of planning policy in the UK' (BRE Global, 2011).

This scenario is not unrealistic. Since its inception, the Code for Sustainable Homes had been a key component in the UK government's strategy to reduce carbon emissions and has been established as a condition of funding in all new affordable housing schemes built in the UK. The latest addition to the BREEAM family, BREEAM Communities, was published early in 2011. In June 2011, Bristol City Council published a Core Strategy as part of their Local Development Framework. This strategy makes it mandatory for all large urban design schemes in the local authority to be assessed using the latest BREEAM Communities form of assessment (Bristol City Council, 2011; BRE, 2012). If such efforts are judged successful, it is possible that other local authorities will follow a similar path.

Yet, this support for BRE forms of assessment is not shared across all departments and scales of government. In July 2010, the Department for Education launched a comprehensive review of all capital investments. This review questioned whether it should be mandatory for each new school over £2 million to achieve a 'very good' rating using the BREEAM New Construction form of assessment (DfE, 2011). Removing this condition, it was suggested, could streamline procurement and reduce additional costs. However, following the passionate and immediate defence offered by representatives of the Chartered Institute of Building, the UK Green Building Council and the Aldersgate group (The Guardian, 2012), these proposals were withdrawn.

Similarly, in 2012 the Department for Local Government and Communities began a much broader and extensive review of regulation and assessment policy through the Housing Standards Review, launched in October 2012 and released in August 2013. The working group suggested a number of changes for subsequent consultation. One of the most significant of these changes was the proposed 'wind down' of the Code for Sustainable Homes and measures taken to integrate these standards into the UK building regulations (DCLG, 2013; 2014). Whilst this change has affected domestic assessments, it has not impacted on non-domestic assessments. Thus, despite these changes to the Code for Sustainable

Homes, the BREEAM forms of assessment remain central to the UK's agenda on sustainability.

This literature shows fierce support for the modelled concept-based assessments used by the BRE assessments as well as wider motivations to challenge and re-think their use in practice. Given the intensity of debate surrounding these assessments, I believe they offer a critical case in which to question the normative use of modelled concepts and to consider whether an alternative is both possible and practicable. If Deleuze can contribute to this alternative, then it may help us re-think the expanding influence of the modelled concept and associated tools.

Researcher's experience using this modelled concept-based tool: whilst not used as part of my selection criteria, I have also taken into consideration the advantages offered by tools in which I, as the researcher, have already developed a strong working knowledge. In Chapter 1, I set out one of many examples taken from my experience and showed how my concerns align with other professionals (see Schweber, 2013). This suggests that my experiences and views do reflect the experiences and views of other experts in the field.

Two research stages and three possible outcomes

The resulting case study of BRE assessments was structured into two research stages corresponding with the two research questions identified at the beginning of this monograph. Possible outcomes for these research stages were identified and captured in the following table (see Table 4.1).

This table presents the two research stages and three possible outcomes. This research strategy was used to guide the theoretical and empirical stages of this study as presented in detail over the following chapters.

An epistemological and methodological approach

So what methods are most appropriate to these two research stages?

For Deleuze and Guattari, a line of enquiry should inevitably lead to an appropriate method. They capture this point in *A Thousand Plateaus*, where they note that 'there is no difference between what a book talks about and how it is made' (Deleuze and Guattari, 2004a). Whilst expressed differently, contributions to research methods literature draw similar conclusions. Key figures in this field have argued that researchers' methods for exploring a field cannot be separated from their research aims and ontological stances (Guba and Lincoln, 1994; Lincoln et al., 2011). Building on this argument, the social scientists Guba and Lincoln categorise different areas of social research according to their ontological, epistemological and methodological positions (Guba and Lincoln, 1994). The combination of these three elements, they argue, forms a paradigm traceable across many different forms of enquiry.

In their attempts to categorise these paradigms, Guba and Lincoln describe and discuss five such groups: positivism, post-positivism, critical theory, constructivism and participatory inquiry (Lincoln et al., 2011: 100). Each of these paradigms

Table 4.1 Two research stages for the study and three possible outcomes

	Research Stage A: *Can Deleuze's concepts be developed into a proposal for a non-essentialist assessment tool?*	*Research Stage B:* *Is there any empirical evidence to suggest that such a tool could be made practicable?*	*Conclusions*
1st outcome	Yes: concepts do translate into a proposed tool.	Yes: concepts underpinning the proposal are considered as a practical direction for assessment practice.	Deleuze's philosophy *can* be used to develop an alternative to the essentialist modelled concepts used in formal assessments. Additionally, there is empirical evidence to suggest that this alternative *can* be translated into a practicable tool; albeit this conclusion would be subject to further development.
2nd outcome	Yes: concepts do translate into a proposed tool.	Not in its current form: practitioners agree with concepts underpinning the proposal but feel they are impractical or problematic in practice.	Deleuze's philosophy *can* be used as a starting point for an alternative to essentialist modelled concepts used in formal assessments. But, to translate and develop this for practice, empirical evidence suggests that changes must be made. Some of these changes may clash with Deleuze's philosophy.
3rd outcome	No: concepts do not translate into a proposed tool.	No: there is little evidence to suggest that a Deleuze-inspired tool would be practicable or useful to practitioners.	Deleuze's philosophy cannot be used to inform an analytical alternative to the essentialist modelled concept. Or it can be used as an analytical alternative, but empirical evidence suggests that it is unable to offer a practicable alternative.

is discussed according to its unique approach to realism or relativism/constructivism[1] and a list of methods typical to these approaches (Lincoln et al., 2011: 100).

These generalities are intended as guidance rather than a definitive reflection of social science research. However, these categories offer very little direction for researchers drawing on Deleuze's philosophy and looking for an appropriate methodology. This is because Deleuze's ontology, epistemology and method do not 'fit' into the five paradigm categories. His actual/virtual ontology and the processes of actualisation that connect these realms (see Chapter 2) cannot be described as 'naïve', 'critical'[2] or 'historical' forms of realism. And whilst Deleuze rejects the idea that we, as researchers, can obtain a true knowledge of reality, his philosophy also argues against relativist, socially constructed understandings of the world around us (DeLanda, 2006).

Whilst social science research methods literature does not provide a clear and well-trodden path between Deleuze's ontology, epistemology, methodology and method selection, a number of Deleuzian scholars offer some guidance for making this connection. Bell suggests that Deleuze's ontology can be structured according to two components: 'pragmatics' and 'transcendental empiricism' (2006). The first of these components concerns Deleuze's approach to creating concepts and developing a line of enquiry. Deleuze draws on concepts from a range of different disciplines and re-creates these concepts by focusing on what they *do* and what they *might do* if they were adapted to different settings and problems (Hillier and Abrahams, 2013). The second of these components, transcendental empiricism, sets out Deleuze's stance on the relation between these concepts and lived experience (the empirical). Deleuze uses this stance to develop and 'test' the pragmatic usefulness of the concepts he creates.

These two components in Deleuze's epistemology correspond with the two stages of my research design: Research Stage A as the construction of a new tool by adapting concepts to a new context (Deleuzian pragmatism) and Research Stage B as the testing of this new tool and the concepts used to create this tool through empirical enquiry (transcendental empiricism). As the following chapters demonstrate, this connection was fundamental to the creation and application of a research design suited to the two research stages.

Notes

1 Positivism as 'naïve realism', postpositivism as 'critical realism', critical theory as 'historical realism', constructivism as relativism and participation as participatory reality (i.e. reality constructed through the participatory project).

2 Note that the ontological and epistemological links between critical realism and a Deleuzian form of realism is discussed in Chapter 6.

5 Methodology for Research Stage A

A pragmatic method for developing a non-essentialist assessment tool

In Chapter 2 I showed how Deleuze's pragmatic approach to concept development relies on the fundamental questions 'what do these concepts do?' and 'what might these concepts do in this specific context?' In Chapter 3, I demonstrated how this pragmatic approach to concepts played out in three Deleuze-inspired studies. In my review of these studies, I highlighted several gaps in their respective frameworks and suggested that such gaps were owing to the researcher's failure to re-create Deleuze's concepts to suit the specific demands of the field. The research design for Stage A was informed by these lessons and by the methodological stages of Deleuze and Guattari's pragmatic method outlined in *A Thousand Plateaus* (Deleuze and Guattari, 2004b).

Deleuze and Guattari suggest that pragmatics, or 'cartography', comprises of four components. The first 'generative' component is described as a method of 'tracing' in which one 'shows how a form of expression located on the language stratum always appeals to several regimes' (Deleuze and Guattari, 2004b: 160). In the context of this book, the process of tracing suggests taking concepts, ideas or terms used in the BRE modelled concept-based assessment tool and showing how it might also appeal to other lines of development, such as those set out by Deleuze-inspired theorists in Chapter 3.

For Deleuze and Guattari, this initial tracing exercise provides the basis for the second 'transformative' component intended to 'show how one abstract regime can be translated, transformed into another, and especially how it can be created from other regimes' (Deleuze and Guattari, 2004a: 161). In order to make this transformation, Deleuze and Guattari suggest that one should form a map of (BRE) concepts and ideas and identify 'possibilities for translat(ing) and creat(ing)' them into new lines of enquiry (into new non-essentialist tools).

Deleuze and Guattari suggest that these two initial stages form the basis of the third component, the 'diagrammatic component', which 'consists in taking regimes of signs or forms of expression and extracting (concepts) from them . . . (such that they are) capable of combining with one another' (Deleuze and Guattari, 2004b: 161). In the context of this study, this process of 'diagramming' would suggest linking the concepts and ideas used in assessment methodologies with the concepts set out in Deleuze's philosophy. This process results in one or several diagrams that show what potentials can and cannot be developed/emerge in the future.

If one is to effectuate these abstract diagrams (abstract machines) into a concrete machine (a practicable assessment tool), Deleuze and Guattari argue that one must explore the fourth 'machinic' component of pragmatism. This 'machinic' component aims to 'outlin(e) the program' for the assessment by showing how it will play out in practice and how it will 'distribute everything' (Deleuze and Guattari, 2004b: 161). In the context of this study, this suggests setting out the Deleuzian ideas underpinning the proposed non-essentialist tool into terms and functions that can be understood by practicing planners, assessors, masterplanners and architects.

These proposals for a four-stage pragmatic method can be seen in the four parts forming Research Stage A (Table 5.1): from the tracing of BRE methods and

Table 5.1 Breakdown of Research Stage A into four sub-stages

Research Stage A: *Can Deleuze's concepts be developed into a proposal for a non-essentialist assessment tool?*			
Research Stage A1: How might we review and (re)think BRE forms of assessment?	*Research Stage A2: Can these ideas be developed into a theoretical proposal?*	*Research Stage A3: Can this proposal be developed by drawing on other forms of assessment used in this milieu?*	*Research Stage A4: How does this translate into a practical proposal?*
The first sub-stage reviewed BRE assessments in order to: • identify and develop initiatives within the BRE's history that might usefully contribute to an alternative. • consider how Deleuze's concepts might re-work these initiatives in the absence of a modelled concept structure.	The second sub-stage used the ideas developed in the last sub-stage (A1) and adapted some of the ideas identified and developed in Chapter 2 and 3 to set out a new method which I have termed 'the Speculative and Immanent Assessment Method' (SIAM).	The third sub-stage responded to some of the practical questions left open in the theoretical proposals for SIAM. It did this by drawing on the established form of assessment used as part of the Construction and Design Management (CDM) regulations. In Chapter 6, I show why the CDM provides useful insight into the gaps left unresolved in the developing SIAM method.	The fourth and final sub-stage 'translated' the theoretical proposals for SIAM into a practical proposal that could be understood and operationalised by practitioners whose knowledge of Deleuze's philosophy might be very limited.

the use of modelled concepts within these methods (Research Stage A1); to the mapping of opportunities for transformation into a non-essentialist tool (Research Stage A1); to mapping the diagram particular to both tools (Research Stage A2); to outlining the resulting assemblage and showing how it might be effectuated into practice (Research Stage A3 and A4).

Table 5.1 presents the process I used to develop and adapt Deleuze's concepts to meet the specific demands of a field of knowledge. The first of these stages acted as a 'scoping' exercise intended to identify opportunities to introduce Deleuze's ideas into the field. This stage was conducted as a documentary review and analysis.

The second and third sub-stages shown in Table 5.1 expanded some of the ideas generated from the historical study and explored other opportunities for introducing Deleuze's concepts into the field of assessment. Much of sub-stage A3 focused on links between the developing proposals for SIAM and the CDM assessments. This assessment was selected following a brief review of formal building and urban design assessments and supported by my own experiences working within this field. This connection between SIAM and the CDM assessment method proved useful, in terms of both the theoretical development of my proposal and the subsequent testing of SIAM through the second empirical stage (Research Stage B). In terms of the latter, it helped direct sample selection and helped explain some of the most difficult Deleuze-inspired ideas (speculation and immanence) into terms that were already understood by designers and regulators alike.

6 Theoretical experiments

Research Stage A

Overview of the chapter

In this chapter I re-trace the process undertaken according to the sub-stages identified in Table 5.1 and use these results to reflect on the overarching question directing this stage of the study: *can Deleuze's philosophical concepts be translated into a new theoretical framework for constructing formal assessments?*

Research Stage A1: reviewing and (re)thinking BRE forms of assessment

In 1990 the Building Research Establishment (BRE) set out their first suite of sustainable assessment methods. These focused on assessing the sustainability of office designs, supermarkets and new-build houses. Eight years later, these assessments were superseded by the introduction of the first BREEAM suite covering a much broader selection of building types. This new group of assessments was distinguished by nine categories deemed essential to 'a sustainable design'. Whilst they have been modified over the years, this basic modelled concept for the 'sustainable scheme' has not changed in any fundamental way over the subsequent seventeen years. Thus, if one were to draw a diagram of all BREEAM assessments formed over this period, one would struggle to distinguish them from the diagrams of the Code for Sustainable Homes and BREEAM New Construction shown in Chapter 2. Whilst this demonstrates the strength of the BRE's conviction in essentialism, one can identify a number of changes in the BRE assessments that suggest opportunities for moving away from this so-called 'universal model' (BRE, 2011). The most striking of these changes are the following:

1 the introduction of BREEAM Bespoke in 2001;
2 the emphasis on flexibility in BREEAM Communities 2012; and
3 the emphasis on community consultation in BREEAM Communities 2012.

The first part of this chapter will focus on these three developments and, where possible, enrol the help of the three Deleuze-inspired theorists discussed in Chapter 3 to ask:

1 what do these aspects do in current BREEAM assessments?
2 what might these aspects do if they were not set within a modelled concept structure?

The introduction of BREEAM Bespoke

What does the BREEAM Bespoke method do?

The BREEAM Bespoke form of assessment was introduced in 2001. It was intended to assess buildings whose uses and contexts were not easily accommodated using existing BREEAM forms of assessment. BREEAM Bespoke filled this gap by adapting the modelled concepts used in other assessments. In each bespoke assessment, a modelled concept is selected from the established suite, traits are added or removed, and weightings are adjusted. Thus, in terms of what the Bespoke assessment *does*, it ensures that established modelled concepts are made applicable to *all* projects including unique and complex projects. More broadly it ensures that all buildings regardless of their use and context can be compared against each other according to the BRE's universal assessment scale, and thus can be judged and regulated accordingly. As discussed in Chapter 1, this approach to assessment reflects Fischer's argument about the importance of systematic assessment as part of a national agenda (2003).

What might a Bespoke assessment do in the absence of a modelled concept?

The idea of a Bespoke assessment that is not structured by a broader modelled concept can be seen in two ways: by looking at Bespoke assessments made by actors as part of the design process, and a Bespoke assessment made by an assessor as part of an evaluative process. The three Deleuze-inspired studies reviewed in Chapter 3 offer insight into these two directions.

Hillier's concept of the 'empty signifier' suggests that the concept of the 'sustainable home' has no inherent meaning. The meaning of the concept changes throughout the design process. So when one is designing the layout of a home, the sustainability of the design might be based on the way the rooms are used and how they might be adapted to accommodate different uses in a changing family environment. When one is designing a cladding detail, one might think about sustainability as the expected life of the materials and how much maintenance is needed across that life. The concept of sustainability, therefore, is part of the immanent process of design: it is dependent on the entities that are brought together at different stages of the design process. It is in this sense that Hillier might discuss the idea of a bespoke *immanent* assessment of sustainability.

A similar position might be developed from concepts introduced in Bonta's theoretical framework. My review of this framework would suggest that the 'space' in which the design is to be introduced, and the broader space of the project, is a *complex space*: a combination of assemblages formed of entities each with their unique operational demands. Because all complex spaces are different, an entity (human or non-human) must assess the complex combination of assemblages around them in order to determine how they should *striate* or *smooth* space. If this is the case, then it is questionable whether one should then assess these striations and smoothness by referring to a set of criteria located outside of the process.

To develop these ideas further, one might argue that all non-essentialist assessments should be seen as bespoke forms of assessment made *immanent within* rather than *transcendent to* (outside of) a complex space. In other words, as I noted in my own experiences in Chapter 1, assessment is already part of the design process, and this should form the basis for assessing a building rather than introducing a separate, or partially separate, evaluative exercise.

Whilst Hillier's and Bonta's (re)created and expanded Deleuzian concepts can be used to consider a bespoke form of sustainable assessment *within* the design process, Halsey's study provides some direction for a bespoke sustainable assessment used to evaluate completed designs. Building on Halsey's naming machines to include non-textual machines (Bonta, 2008; Abrahams, 2013), one might assess a building or a masterplan by asking, 'what becomings are blocked or encouraged as a consequence of the design in this instance?' Thus, assessing the sustainability of a scheme would require efforts to consider how the design encouraged or blocked the 'becoming sustainable' of entities within a complex space (Bonta and Protevi, 2004).

This offers two Deleuze-inspired directions for developing the concept of a Bespoke sustainable assessment in the absence of a modelled concept. These are:

- as part of the design process within a complex space; and
- through the evaluation of that design and how it affects a complex space.

An emphasis on flexibility

What does flexibility do?

The technical manual for the 2012 Community assessment emphasises its flexibility (BRE, 2012). The BRE considers flexibility important because it 'reflect(s) the unique challenges and opportunities on different development sites' (BRE, 2012: 2). Whilst flexibility was a key part of former assessments, this was mostly achieved by offering designers and assessors different criteria for achieving a given Code or BREEAM level. Yet, in the Communities assessment, most of the criteria used in two of the three assessment stages are optional. The only mandatory criteria are those belonging to the initial scheme design stage and the criteria for consultation with stakeholders in subsequent design stages.

By allowing greater degrees of flexibility, the Communities assessment draws the BRE much further from their essentialist ideals than the Bespoke form of assessment. It suggests that many of the seven categories and the thirty-nine criteria within those categories are not essential to all projects. As with the Bespoke form of assessment, this focus on flexibility serves a very specific role in the assessment. It allows the overarching structure of the modelled concept to be used to assess greater degrees of complexity.

What might flexibility do in the absence of a modelled concept?

Hillier's (re)creation of the Deleuzian concept, 'plane of immanence', suggests that the early strategic design phase is the point at which actors draw becomings from the virtual plane of immanence into a figure of the virtual plane (a strategic plan). It is thus at this early strategic stage that actors speculate what *might become* of the actual world around them. The Communities form of BREEAM assessment suggests that this stage should include a number of mandatory steps, captured in the mandatory aspects of the assessment criteria. By basing these speculations on pre-conceived design stages and using these as mandatory criteria in subsequent assessment, Hillier might argue that these actors are using a blueprint (*plane of transcendence*) to set out a strategy (*a plan[e] of immanence*) and to assess this strategy. And, as a result, this strategy is limited to what is already known, or what has already been identified and assessed in other projects with definable similarities.

Expanding on the Deleuze-inspired concepts developed by Hillier, a flexible, sustainable assessment would be one that allows for the greatest number of speculations: one that allows actors to stretch as far as possible 'beyond the horizon' of what we already know and see (2007). According to Hillier, this can be achieved only by removing these blueprints as starting points.

These concepts, therefore, provide a direction for developing a flexible sustainable form of assessment in the absence of a modelled concept. This is rooted in:

- a design process that allows the greatest degree of speculation about 'what might become' during the earliest strategic stages of a design and an assessment that encourages such speculations.

An emphasis on community consultation

What does community consultation do?

BREEAM Community 2012 is the first assessment to emphasise the importance of community consultation in assessing the sustainability of a design. Despite its importance, community consultation has a limited role to play in BREEAM Communities. Consultation is deemed an integral part of the design process but not the process of assessment. In other words, the assessment emphasises the importance of stakeholder engagement to resolve design problems arising in different contexts but not to assess the results of these decisions. This method remains under the control of the BRE. In doing so, the BRE form a clear distinction between design and assessment.

What might community consultation do in the absence of a modelled concept?

If one were to explore how the concept of consultation might be expanded in the absence of this structure, one might consider what would happen if the means of forming an assessment was also determined through consultation. In terms of the concepts discussed and expanded from the literature review, this would mean integrating design-based assessments of sustainability drawn from Hillier's and Bonta's frameworks and evaluative assessments of sustainability drawn from Halsey's framework.

To explore this idea, I started by considering Hillier's position on community consultation through 'consensus building'. As noted in Chapter 3, Hillier rejects the idea that consultation between different actors should conclude in consensus. In the same way that a flexible assessment should not start with a blueprint that determines what should be speculated, she notes that consultation should not start with pre-defined ideals about what consultation is supposed to achieve.

Developing this idea in the context of design and assessment, consultation in both processes would be seen as a form of 'bricolage'. This would mean assessing the sustainability of a project as the project team moves through the design process. In the same way that consensus is removed as an end-point in consultation, it would suggest removing the idea of a pre-determined Code or BREEAM level that the design must achieve and a scale by which all projects can be compared – i.e. what Hillier termed the 'transvaluative' approach to consultation (2007).

This transvaluative approach offers other advantages. Without a known end-point, it would be difficult for actors to predict what the implications of this immanent assessment process might be. As a result, the decisions and the assessment of decisions would be based on the group of entities under consideration at any given point in time rather than thinking about how those decisions might affect a final sustainable score.

This presents a direction for developing a sustainable assessment method based on community consultation in the absence of a modelled concept, or a group like the BRE enforcing a modelled concept. This direction suggests:

- combining the assessments made as part of the design process and the assessments made to evaluate this design;
- consulting with the same actors in both design and evaluation assessments; and
- constructing this assessment in a way that avoids a pre-conceived end point, in terms of either what the design will become or what the evaluation of that design might mean.

This first part of Research Stage A shows how concepts taken from the BREEAM assessments might be developed differently if they were developed through Deleuze's ontological position rather than the essentialist stance underpinning modelled concepts. By drawing on and expanding some of the ideas from other Deleuze-inspired theorists I have set out a number of assertions about the way modelled concepts affect the way one thinks about bespoke assessment, flexibility and public consultation. But I have also shown how these same ideas might be developed differently if they were not constrained by the modelled concept framework.

Research Stage A2: developing these ideas as a summary of the BRE method and an outline for a theoretical proposal

Up to this point in the book, I have presented, discussed and elaborated on a number of Deleuzian concepts and Deleuze-inspired ideas. I have shown how these concepts can be used to explain the design process (Chapter 2), to analyse and help us engage in a range of different spatial issues (Chapter 3) and to develop some of the concepts used in BREEAM to inform a new non-essentialist assessment method (this chapter).

The aim of this part of the chapter is to bring these ideas together to provide a summary of the BRE assessment method from a Deleuzian perspective and to present my proposals for a new Deleuze-inspired method, which I have termed 'the Speculative and Immanent Assessment Method' (SIAM).

Deleuze's concepts and a summary of the BRE Environment Assessment Method (BREEAM)

Objects and their properties

According to BRE forms of assessments, a building or city can be assessed as an object according to its discernible properties. In doing so, this approach to assessment does not consider how the design came to have these properties, the decisions that were made and the decisions that might be made in subsequent stages. In this sense, the point at which an assessment is undertaken, and the object that is assessed, is considered *transcendental* to the processes of actualisation.

As I noted in Chapter 2, this approach is made possible because of the essentialist stance taken by the BRE, and the *relations of interiority* that underpin essentialism (DeLanda, 2002). As DeLanda notes, relations of interiority assume that a whole can be understood according to the cumulative properties of its components. This applies to all material and non-material entities and to different aspects of a building or city, including the processes of design, the spaces that form this design, the materials that form these spaces and the methods used to assess the design. A final building design is considered as a collection of partial stages, as captured in the official stages of work, set out by the Royal Institute of British Architects for example. The BRE use these stages to determine when an assessment should be undertaken. The collation of these stage-based assessments is then used to form conclusions about the building as a whole.

Similarly, at each of these component stages, a building can be understood as a whole, formed from its components such as its rooms, its walls, the roof, its foundations and so forth. All of these components can, in turn, be broken down into their components. Thus, the wall is considered as a whole, formed from bricks, blockwork, brick ties, insulation, plasterboard, screws, wet plaster skim. As shown in Figures 2.1 and 2.2 (Chapter 2), the BRE use these material components as criteria defining an essential trait of the sustainable home, building or community. This shows how the relations of interiority are related across different aspects to form a 'holistic' assessment.

Referential model

Relations of interiority explain the construction of assessments based around a modelled concept, whilst Deleuze's concept of the *referential model* helps explain the role this modelled concept plays in processes of design and assessment. By linking across different aspects of a building design, this referential model determines what a sustainable home, building or community is; what it should be made of; and how it should be designed. Because all buildings and urban schemes are judged against this model, any variation is considered as *differences from* the model. Likewise, all similarities are judged to be *repetitions of* this model.

End points

The modelled concept serves as a preconceived, transcendental *end point*, broken down to form a value system. In BREEAM assessments, this is the rating of a building as pass, good, very good, excellent. The decision to reach a minimum rating on this value system is determined before most of the design processes begins and, thus, before assessing what entities form a site and project, or what the operational demands of these entities are or might become.

Deleuze's concepts and a proposal for a Speculative and Immanent Assessment Method (SIAM)

Assemblages

A Deleuze-inspired, non-essentialist alternative form of assessment would approach the building and its actualisation in a very different way. Once constructed, a building or a city would be understood as an *actualised assemblage*. Drawing on Deleuze's realist stance, this building or city assemblage would be treated as one of many assemblages forming the actual world around us. In this respect, a building or city assemblage has no ontological distinction from any other actualised assemblage and cannot be separated from them. This is because all assemblages are actualised according to *relations of exteriority*. In simple terms, this means that an assemblage operates in ways that are more than the summation of its parts. If we were asked, 'what does a house assemblage *do*?' we would gain little insight by listing the walls, bricks and slates that form the building, the people who live in the building or the assessments used to evaluate the building. These components do not explain how the building affects regeneration, the movement of people through a city, the effects on tourism, or the movement of water through the ground or birds through the sky. Similarly, the entities that form the building assemblage (can) do more than their role in the assemblage.

Becomings and blocks of becoming

Rather than focusing on a modelled concept acting as a *referential model*, a Deleuzian assessment would focus on the processes of actualisation or, in this instance, the design process. In Chapter 2, I used Deleuze's ontological concepts of *becoming* and *blocks of becoming* to help make sense of this complex process.

I showed how, in simple terms, the design process is formed from a number of *blocks of becoming*, each based on, what architects might term, a 'design problem'.

Such blocks might include a 'becoming structurally stable of the design', 'becoming corrosion resistant of the steel', 'becoming concrete of the frame' and so forth.

To develop these blocks of becoming, a design team experiments with the entities in a developing building or urban design assemblage, as well as entities found in other connected assemblages, such as manufacturing assemblages and transportation assemblages for example. Each of these experiments demands that the project team *speculate* and *assess* how such entities might help resolve the 'design problem' under consideration, or, in Deleuzian terms, how their *becomings* contribute to a broader *block of becoming*. As I demonstrated in Chapter 2, there are some instances in which these assessments might lead to significant changes in the design by re-directing the design process along a new block of becoming. In other instances this process of speculation and assessment might identify new 'design problems', which also result in a new set of experiments directed along a new block of becoming.

To assess this process in terms of sustainability, a Deleuzian assessment would be positioned *within* the design process itself. This *immanent* form of assessment would be undertaken by members of the design team rather than an external assessor. As in the design process itself, members of the design team would be encouraged to speculate how the entities they bring together, and the experiments they develop, might have the potential to become sustainable. In this sense, this assessment acts as a broad *becoming sustainable* block running through the entire design process.

Unlike essentialist forms of assessment, the concept of sustainability is used as an *empty signifier*. As such, its meaning is not determined in advance but rather is derived from within the process itself. To illustrate this immanent and speculative form of environmental assessment, I return again to the extracts from practice discussed in Chapters 1 and 2. Drawing on this example, I noted that the design team had decided to change the design from a steel structural frame to a concrete structural frame in response to a number of factors, including the corrosive tendencies of airborne salt from the sea. Because the building was assessed using a modelled concept-based environmental assessment (the Code for Sustainable Homes), this decision was deemed unsustainable, owing to the higher CO_2 emissions of concrete over steel.

Using an immanent and speculative form of assessment, CO_2 emissions would be seen as a potential 'to become sustainable'. However, it would not be seen as the only potential to become sustainable within this area of the design. Other becomings would be drawn from factors taken into account during the design process. Thus, whilst CO_2 emissions offered a potential to 'become sustainable' blocked by the introduction of concrete, this same decision encouraged other potentials to 'become sustainable', such as those associated with reduced maintenance (necessary if the steel was to resist the long-term corrosive tendencies of airborne salt); reduced depths of insulation (owing to the improved thermal properties of concrete over steel); reduced contractor programme, road closures and the number of tradespersons working on site (owing to a structural design that could support poor ground stability as well as an above-ground building structure). All these potentials to become sustainable offer different meanings of the concept. Taking this into consideration, an immanent and speculative form of environmental assessment might have concluded that the decision to use a concrete frame rather than a steel frame encouraged more sustainable becomings than it blocked, thus judging this decision favourably.

A project team's capacity to identify and assess becomings, as part of either a design process or a sustainable assessment process, is limited by its imagination. For a Deleuzian assessment to operate effectively, it is important that every effort is made to encourage these actors to 'stretch' their imagination as far away from the actual as possible, or, what Hillier termed, to 'stretch beyond the horizon'. But there will always be a number of potentials (becomings) that cannot be imagined: what one might term 'un-speculatable potentials'.[1] These potentials emerge during the process of actualisation. In the same example, ground surveys revealed deep sand strata below the site, aspects of the design became the focus of local concern resulting in a number of planning objections, the price of steel rose considerably between the initial sketch design and the detailed planning application, and a report was published that highlighted the damaging effects caused by seagulls to rubber-based roofing membranes used on coastal buildings (Wells, 2007). Each of these events emerged during the design process and revealed potentials that affected the design. Thus, an immanent and speculative form of sustainable assessment would need to assess how the design responded to these un-speculatable potentials by asking what sustainable becomings were encouraged or blocked as a consequence of the emerging event and the corresponding design change.

This brief description illuminates the differences between a modelled concept-based form of assessment (BREEAM) and a Deleuze-inspired form of assessment (SIAM). This distinction is captured in Table 6.1 below.

Table 6.1 summarises some of the points raised above concerning the distinction between BRE assessments and SIAM. The table is separated into two sections. The first section, 'ontological framework', captures some of the concepts drawn, expanded and (re)created from Deleuze's philosophy to suit the unique demands of sustainable assessment, whilst the second section starts to think about how these ideas might play out in practice. This transition is important because it highlights the key issue guiding this book: the practical usefulness of Deleuze's philosophy in planning. If we look along this second section, this table shows how I used Deleuze's ontological framework to create a new assessment tool. However, the final row highlights two questions that must be resolved if this tool is to be made viable:

- how would SIAM be introduced into current practice?
- how would SIAM ensure that the project team prioritised speculations for becoming sustainable in the design process?

To answer these questions, and to expand and develop Table 6.1, I explored another area of assessment practice affecting the way architects, masterplanners and planners form design decisions. In the UK, all project designs are subject

Table 6.1 Distinction between the BRE assessment method and the proposed Speculative and Immanent Assessment Method (SIAM)

	BREEAM	*SIAM*
Ontological framework	Essentialist	Pragmatist (in a Deleuzian sense)
	Modelled concept of 'the sustainable home, office' etc.	A block of becoming 'sustainable', where sustainability is an empty signifier
	Transcendental process	Immanent process
	Referential model	Speculated potentials
	Building as a whole defined by its properties	Building as assemblage formed of actualised potentials
Assessment in practice	Assessment process is separate or parallel with design process	Assessment is part of/inseparable from design process
	Assessment undertaken at pre-determined stages of the design	No pre-determined stages for assessment. Assessment undertaken when member of the project team feels a key or difficult design decision is made during the design process
	Assessment by external assessor	Assessment by members of the project team
	Assessment results in a score judged against a universal value system (valuative)	Assessment results in an on-going record of key design decisions, the entities that were instrumental to that design decision and how the decision encouraged or blocked their potentials to be become sustainable (transvaluative)
	Sustainability score acts as a benchmark that the design must achieve	There is no benchmark; each decision is judged as more or less sustainable according to the complex factors that contributed to the decisions taken during the process
	One sustainability score is used to explain the sustainability of the scheme as a whole	No attempt to explain the sustainability of the design as a whole; rather, a series of connected assessments that show how a building was actualised through a series of decisions judged as more or less sustainable
	The score as benchmark is enforced as a condition of planning approval or a condition of a client's brief; this ensures that the project team works in ways that are deemed 'sustainable' (through the essentialist understanding of the term)	*How would SIAM be introduced into current practice?* *How would SIAM ensure that the project team prioritised speculations for becoming sustainable in the design process?*

to Health and Safety assessments as set out by the Health and Safety Executive (HSE) and defined by the Construction and Design Management regulations (CDM) (HSE, 2007; 2015).

Unlike BREEAM, this regulatory framework is not constructed around an essentialist definition, such as the 'safe design'. Indeed, as in my proposed SIAM, the CDM regulations suggest that assessment should be undertaken as part of/ within the design process, through *speculations* about what might become of the design if constructed. In this sense, the CDM form of assessment (in both 2007 and 2015 iterations) is much closer to the speculative Deleuze-inspired approach to assessment outlined above.

Research Stage A3: developing the proposal by drawing on other forms of assessment used in the design process

The first CDM regulations were introduced in 1995 in response to the EU directive 92/57/EEC on the implementation of minimum safety and health requirements at temporary or mobile construction sites (CEC, 1992; HSE, 1994). These regulations intended to reduce deaths and injuries during the construction and management process by changing the procedures used by associated professionals. The CDM regulations were revised in 2007 (HSE, 2007) and revised further in 2015 (HSE, 2015). The 2015 revision reduces the perceived bureaucracy surrounding CDM assessments by allowing members of the design team to take on the role formerly known as the CDM Coordinator, now termed 'the principle designer'. This change ensures closer links with EU directive 92/57/EEC, as set out in the Lofstedt report (Lofstedt, 2011).

In the following, I will provide a brief overview of CDM regulations drawing on the established 2007 iteration and any key variations noted in the more recent 2015 iteration using the three points identified above as sub-headings. As part of this review, I will discuss the compatibility of these points against a Deleuze-inspired theoretical framework and as part of a practicable assessment method. The results of this exercise will be used to consider which, if any, of these ideas could be usefully incorporated into SIAM, and what the implications of this would be in terms of Deleuze's usefulness to forming a viable tool for practice. The resulting proposal will be used to expand and develop Table 6.1 above (see Table 6.2).

Assessment during the design process

The CDM regulations 'integrate health and safety into the management of the project' in terms of management of both the design process and the construction process (HSE, 2007: 7) – the intention of which is

> to focus attention on planning and management *throughout* construction projects, from design concept onwards . . . to ensure that health and safety issues are identified, integrated into the overall design process and addressed as they go along.
>
> (HSE, 2007: 7; 32 emphasis added).

Whilst approached from a practice-based rather than a theoretical angle, this proposal for a form of assessment undertaken as part of and within the design process, or what Deleuze might have termed an 'immanent' form of assessment, reflects the broad objectives captured in SIAM.

Speculations

Under the CDM regulations, all designers have a legal duty to ensure that every effort is made to reduce foreseeable health and safety risks in their proposed designs. There are no formal mechanisms used to enforce this obligation. As the HSE note:

> risk assessment of a design should be integral to, and evolve with, the design work itself. Every design is different, and every design will require a degree of calculation, assessment, review and the proper exercise of judgement.
>
> (HSE, no date)

These risk assessments are constructed for a specific project or for a specific practice and are intended to help the designer speculate foreseeable risks and measures to reduce these risks. This idea of the foreseeable versus the unforeseeable risk is captured in the regulations which note:

> Designers are required to avoid foreseeable risks 'so far as is reasonably practicable, taking due account of other relevant design considerations'. The greater the risk, the greater the weight that must be given to eliminating or reducing it. Designers are not expected to consider or address risks which cannot be foreseen, and the Regulations do not require zero risk designs because this is simply impossible.
>
> (HSE, 2007: 32)

Whilst expressed differently, this reference to foreseeable and unforeseeable risks captures the two areas of speculated potential discussed by Hillier – i.e. the potentials that one can imagine (speculated potentials) and those that one cannot imagine (unspeculated potentials). As I argued above, the latter occurs when the design team does not have sufficient information about assemblages in the project to make these speculations, or because they were not accounted for in the early stages of the design.

Another area of compatibility with a Deleuze-inspired approach can be seen in the way these speculations are formed. The HSE suggest that the CDM

> regulations do not prescribe design outcomes [or pre-defined end-points such as those captured in the BREEAM coding system], but they do require designers to weigh the various factors and reach reasoned, professional decisions.
>
> (HSE, 2007: 32)

Whilst the CDM regulations do not rule out the use of referential models entirely, this comment rejects the idea that such referential models should determine when an assessment should be undertaken or how speculations should be formed and judged.

In Deleuzian terms, the CDM regulations suggest that assessment should be seen as a constant process of speculation of potentials to become safe (or unsafe). The decision to capture these speculations as a separate document (i.e. a risk assessment) is determined by the designer. Whilst there is no formal guidance available for making this decision, the assessment process incorporates two mechanisms to ensure that risk assessments are undertaken at key points in the design: by incorporating risk assessments across *all* overlapping design elements, and by introducing a CDM Coordinator (HSE, 2007) or Principle Designer (HSE, 2015) role responsible for planning, managing, monitoring and coordinating health and safety issues. These two mechanisms are discussed below.

Assessment by members of the project team

Under the CDM regulations, the legal responsibility to reduce health and safety risks to the best of their ability and judgement is applicable to all members of the project team including the designers, the client, the contractor and associated building trades. To operationalise this responsibility, all designers are expected to undertake a risk assessment within the limitations of their design proposals and to consider how their proposals might impact on other associated aspects of the designs or management of the buildings. Thus, the architect would assess the risks associated with the design of a particular roof configuration and how this might introduce or mitigate the risks associated with erecting, maintaining or replacing this roof. The architect must also consider how this design affects the risks associated with the structural engineer's proposed design for the steel frame used to support this roof, who will, in turn consider the risks associated with fabricating the frame, bringing it to site and erecting the frame. The structural engineer's risk assessment will also consider other aspects of the design, such as how the proposed steel frame affects the risks associated with the ductwork design proposed by the mechanical and electrical consultant and so forth. Because these overlaps must be discussed between different groups of designers, the assessment of risk and the speculations that form these assessments are subject to constant review within the design team.

A more formal mechanism for identifying key areas of the design for assessment, and the speculations used in this assessment, can be seen in the role played by the CDM Coordinator in the 2007 iteration and the Principle Designer under the latest 2015 iteration (HSE, 2007; 2015).

Unlike the assessor described in BREEAM and the Code for Sustainable Homes, this CDM Coordinator or Principle Designer is described as a member of the project team whose principle tasks are to facilitate co-operation and co-ordination in matters of health and safety in the design process, to liaise with the principle contractor on design matters as they arise and to prepare and update the health and safety file.

During the design process, the CDM Coordinator or Principle Designer encourages members of the design team to speculate potential health and safety risks (becoming unsafe) through all areas of decision making. They work with these designers to identify areas of risk and how such speculated risks can be reduced. They ensure that the risk assessments record these speculated risks, the way the design responds to these risks and why the resulting design is considered the most appropriate solution.

The CDM Coordinators or Principle Designers discuss this developing sequence of assessments with different members of the design teams as well as the contractor teams. In doing so, they can ensure that the resulting design accounts for a range of speculations, in terms of both *what might become* of the building when complete and *what might become* of the building as it is being constructed. To illustrate this process, I would like to draw on one of my experiences from architectural practice.

In 2010, I acted as one of several architects on a large school design. During the detailed design stage, the design team considered a number of options for the external wall construction. The two principle options were masonry construction designed with a brick outer leaf, and a part masonry construction with an external cladding system. To decide which of these was most suitable, we speculated a number of potentials *to become durable* (bricks are more resistant to the wear and tear associated with the building type), *to become structurally stable* (brick construction offers greater lateral stability than cladding panels), *to become cost effective* (bricks are cheaper, in terms of both material and labour costs) and *to become easier and faster to construct* (setting out brickwork on site is easier than setting out cladding panels). In the risk assessment, we focused on potentials *to become safe/unsafe*. We noted the risks that might result if the design required a number of trades working in the same space at the same time (brick layers erecting both leafs rather than combining bricklayers and cladding panel specialists), the reduced demand for heavy machinery to erect brickwork rather than cladding, and the reduced maintenance and thus the reduced risks associated with replacement or repair of brickwork over cladding. After we discussed these options and reasons with the CDM Coordinator, it was agreed that brick was a design solution with a lower safety risk than all other viable options.

However, the CDM Coordinator's consultation with the Principle Contractor introduced other speculations not considered during the design process. Building a large brick wall requires prolonged use of scaffolding. Working from scaffolding increases the risks of falling during construction and the risks of falling from an unattended site (i.e. by possible tress-passers). Measures to reduce these risks also affect other becomings within the design process. They extend the construction programme and limit the ways in which the school can operate during the construction process. Taking these points into consideration, we later agreed that a brick-slip cladding system – i.e. a system that uses small sections of brick fixed to a cold steel frame – would be used. This system could be erected and maintained using cherry-pickers around the perimeter.

This short example shows how the CDM Co-ordinator worked as part of the project team, encouraging different members of that team to speculate potential risks associated with their area of knowledge and to bring these speculations together. Similar observations are identified in broader literature. In interviews with contractors, Atkinson and Westall note how one interviewee believed that using such processes 'allowed the contractor to educate the designers to look at the design from a site safety point of view' (Atkinson and Westall, 2010).

This short overview of CDM forms of assessment shows areas of compatibility with the three ideas used to structure SIAM: a form of assessment based on speculations; a form of assessment made as part of/within the design process (immanent); and a form of assessment made by members of the design team rather than an external assessor. In doing so, the CDM form of assessment provides some directions for responding to the two questions left unanswered above. With this in mind Table 6.2, draws on CDM assessments to expand and develop the comparison

Table 6.2 Developing the distinction between the Building Research Establishment Environmental Assessment Method (BREEAM) and the proposed Speculative and Immanent Assessment Method (SIAM)

	BREEAM	*SIAM*
Ontological framework	Essentialist	Pragmatist (in a Deleuzian sense)
	Modelled concept of 'the sustainable home, office' etc.	A block of becoming 'sustainable', where sustainability is an empty signifier
	Transcendental process	Immanent process
	Referential model	Speculated potentials
	Building as whole object defined by its properties	Building as assemblage formed of actualised potentials
Assessment in practice	Assessment process is separate from design process	Assessment is part of/inseparable from design process
	Assessment undertaken at pre-determined stages of the design	No pre-determined stages for assessment; assessment undertaken when member of the project team feels a key or difficult design decisions is made during the design process
	Assessment by external assessor	Assessment by members of the project team; *all members of the project team undertake an assessment based on their own contribution – the only exception being the addition of a Sustainability Assessment Coordinator whose role is to ensure co-operation and co-ordination between assessments and to encourage designers to prioritise sustainable speculations in their proposals*
	Assessment results in a score judged against a universal value system (valuative)	Assessment results in an on-going record of key design decisions, the entities that were instrumental to that design decision and how the decision encouraged or blocked their potentials to be become sustainable (transvaluative); *these assessments are compiled by the Sustainable Assessment Coordinator*

BREEAM	*SIAM*
Sustainability score acts as a benchmark that the design must achieve	There is no benchmark; each decision is judged as more or less sustainable according to the complex factors that contributed to the decisions taken during its actualisation
One sustainability score is used to explain the sustainability of the scheme as a whole	No attempt to explain the sustainability of the design as a whole; rather, a series of connected assessments that show how a building was actualised through a series of decisions judged as more or less sustainable; *this series of assessments is compiled during the design process*
The score as benchmark is enforced as a condition of planning approval or a condition of a client's brief; this ensures that the project team work in ways that are deemed 'sustainable' (through the essentialist understanding of the term)	*How would SIAM be introduced into current practice?* *How would SIAM ensure that the project team prioritised speculations for becoming sustainable in the design process?* *One possibility is to introduce a formal duty applicable to all members of the project team to ensure that every effort is made to make the most of, or account for, foreseeable potentials to become sustainable. This could be a legal duty as in CDM or a condition of planning.* *There are no formal, referential models that determine how this duty should be met. Thus, SIAM provides only a broad framework open to adaptation. In this instance, all members of the project team would be encouraged to undertake sustainable assessments throughout the project. The role of encouragement and co-ordination would be the responsibility of the Sustainability Assessment Coordinator, who, as in the 2015 iteration of the CDM regulations, would be appointed from within the project team.*

between BREEAM and SIAM. For the sake of clarity, additions to the table are shown as italicised text.

Research Stage A4: setting out a practical proposal

The aim of this final sub-stage is to 'translate' the theoretical proposals for SIAM from Table 6.2 into a practical proposal that could be understood and operationalised by practitioners whose knowledge of Deleuze's philosophy might be very limited. To do this, I set out a brief presentation of SIAM and how it might be operationalised and institutionalised in practice.

1 SIAM is described as an *immanent* form of assessment because it is designed to be part of/within the design process.
2 SIAM recognises that assessments are *already* part of/within the design process. To form a design decision, a designer must identify important factors affecting the scheme and assess their potential contribution.
   SIAM aims to bring these design-based assessments into the foreground of assessment methodology.
3 SIAM recognises that these assessments in the design process are not made in isolation. Assessments based on sustainability must be balanced against other goals such as the 'cost effective building', the 'structurally stable building', the 'innovative building', the 'building that meets user requirements' etc.
   SIAM focuses on one of these aims, 'the sustainable building', but encourages assessors to discuss the way other goals impact on a design decision.
4 SIAM recognises that a sustainable solution in one building may be deemed less sustainable for another. This is because there are different ways of thinking about sustainability: as longevity, CO_2 emissions, maintenance demands etc. SIAM also understands that all materials used in a design make a unique contribution to the scheme and to different understandings of 'the sustainable building/urban design'.
   SIAM is conceived to allow for these different ways of thinking about sustainability and for the different ways in which parts of a design might contribute to these understandings.
5 SIAM recognises that designers do not conceive a building or urban design as a coherent whole but rather as a collection of design decisions that work together.
   SIAM is thus conceived as a collection of assessments made for each key decision as and when they arise.
6 SIAM recognises that these decisions and the assessments that form these decisions are complex and are best understood by the person who made them.
   SIAM aims to avoid any distinction between the designer/decision maker and the assessor. It does this by making each member of the design team responsible for assessing his or her own contribution to the building or urban design. Thus, each designer/decision maker must decide which areas of the design process should be assessed, which factors are important to the assessment and how these factors contributed to a sustainable design.
7 SIAM is offered as a *speculative* method of assessment because it recognises that all design decisions are made speculatively – i.e. about a building or an

urban design that may or may not exist in the future. It recognises that these speculations are made with no way to be sure whether they will materialise and no way of knowing whether all important factors have been accounted for and assessed accordingly.

SIAM aims to capture this speculative approach. It does not assess a building based on the properties of a final proposal. Rather, it aims to assess a design decision based on speculations made by a designer/decision maker concerning potentials to become sustainable and how the designer/decision maker assesses these speculations and uses them to influence the design.

To illustrate how a SIAM assessment might be carried out in practice, the following points sketch out a sequence of events based on a traditional architect's appointment and procurement route as defined by the RIBA. Other variations of this sequence based on other forms of appointment or procurement routes would need to be explored in future research.

8 During the early briefing stages of a design (RIBA stage 1), the lead designer will inform the planning department that it will undertake a SIAM assessment rather than a BREEAM or CSH assessment. This marks the beginning of the assessment process.
9 The client or the lead designer is appointed as the Sustainability Assessment Coordinator. This role may or may not be transferred to a different member of the project team as the project moves through subsequent stages of design development. This may be particularly useful in the detailed design stages during which other members of the team may have a key role in the design process.
10 As the design develops, the Sustainability Assessment Coordinator encourages designers to continuously assess the sustainability of a scheme and balance these against other assessments such as those based on cost, user requirements, health and safety etc. On larger projects, this may be best undertaken through regular design team meetings; on smaller projects, this may be done in-house through regular informal discussions between staff.
11 Regular discussion with the Sustainability Assessment Coordinator helps the designer to decide when an assessment should be undertaken.
12 Such assessments are most beneficial when the designer feels that a design combines and negotiates different definitions of sustainability made possible through the selection of different entities.
13 To form this assessment, the designer is encouraged to follow three rough stages:

Stage 1: identify entities in the project

14 The designer undertakes a 'mapping exercise' of all the entities that constitute the design 'problem'. These will include a range of material entities such as site conditions or other aspects of the building/urban design. But it will also

contain non-material entities such as key areas of the design brief, planning guidance, survey recommendations.

Stage 2: speculation

15 In the second stage of assessment, the designers are encouraged to make a note of their speculations. These will concern questions such as, what do these entities do currently? What else might they do if they were part of this scheme? In what way is this sustainable? How does this affect other goals for the project, such as cost efficiency, meeting user requirements etc.?

Stage 3: discussion

16 In the third stage of assessment, designers are encouraged to discuss the way in which they tackled these speculations: which speculations from stage 2 were taken forward, and which were not? Why? What benefits were thought to result from this decision? How did they affect other areas of the design, and how were these accommodated? And, most importantly, after taking these things into consideration, why do they believe this to be the most sustainable solution for the scheme?

17 This assessment is then discussed with the Sustainability Assessment Coordinator. During this discussion, the Sustainability Assessment Coordinator introduces speculations that have not been considered by the designer. These speculations are drawn from other areas of the design being assessed by other members of the design team. These discussions may result in more or less changes to the design as it develops further.

18 Once agreed in principle, the Sustainability Assessment Coordinator adds the assessment to the sustainability assessment file, which can be accessed by all members of the project team at any given stage in the design process.

19 The sustainability assessment file is discussed in early planning consultation with the planning officer and discussions with the building regulation officer or appointed inspector during detailed design stages. During these consultations, the planning officer or building regulation officer is also asked to speculate potentials and to discuss these with the Sustainability Assessment Coordinator.

20 These discussions are noted in the file and sent to relevant members of the project team for subsequent amendments of the assessment.

21 All sustainability assessments are collated and summarised by the Sustainable Assessment Coordinator. This is submitted as a supporting document in the formal planning application and later as a supporting document in a full plans building regulation submission.

Conclusion to Research Stage A

In my review of the three Deleuze-inspired voices – Bonta, Halsey and Hillier – I suggested that to make Deleuze's philosophy useful, one must be prepared to (re)create his ontological concepts to respond to the specific demands of a given field of interest. Failure to do so, I argued, left gaps and problems in the resulting theoretical framework. This pragmatic approach has served as the methodology for the first of two research stages in this study. The results of this theoretical exercise discussed in this chapter suggest that Deleuze's concepts *can* be adapted to inform a new non-essentialist tool for assessing the sustainability of building and urban designs. This research stage also shows that, to make this transition, one must identify opportunities for adaptation from within the field, within other related fields as well as within broader Deleuzian scholarship.

Note

1 See my review of Hillier's study in Chapter 3.

7 Methodology for Research Stage B

Outline of the chapter

In Chapter 5 I set out a pragmatic method used to develop Deleuze's concepts into a theoretical assessment tool for practice. In this chapter I look to the second part of Deleuze's method, 'transcendental empiricism'. This chapter explains how I used Deleuze's theoretical arguments for transcendental empiricism to form a workable empirical method that helped test my proposal for SIAM.

Epistemology, methodology and research strategy for Research Stage B

Deleuze's epistemological stance, termed 'transcendental empiricism', is built around a critique of established empirical methods. This is captured by the Deleuzian scholar, Levi Bryant, who notes that most forms of empirical enquiry, including those conducted by most pragmatists, share a common interest in the role, relationship and status of objects and subjects. This focus is often defined by questions such as, 'how can we, as subjects know objects or/and subjects?', 'how are subjects produced by objects?' or 'how are objects produced by subjects?' (Bryant, 2008a: 265). This reflects what Bryant terms the 'primacy [often given to] the subject and object' (2008a: 265).

As discussed in Chapter 2, Deleuze holds that subjects and objects are both products of *becomings* drawn from processes of actualisation. As such, empirical enquiry cannot and should not take these products as the starting points for generating a theory (Bryant, 2008a). For Deleuze, if we are to engage in empirical enquiry, we must do so without ignoring or underplaying the importance of the becomings that form the entities we see in the world around us (processes of actualisation) (DeLanda, 2002; Bryant, 2008a: 265). Importantly, whilst Deleuze holds that empirical enquiry cannot ignore these processes, he also holds that we cannot observe them either (Bryant, 2008a). According to Deleuze's stance, we, as researchers, are actualised subjects whose sensory experiences can only identify other actualised entities. In other words, for Deleuze, the pre-actual realm cannot be accessed through empirical enquiry yet must still be accounted for and used to explain our empirical observations. It is in this sense, therefore, that Deleuze's approach to empiricism is considered 'transcendental' (Bryant, 2008a) or 'higher' (Williams, 2008).

This argument opens up a difficult question that I believe is important to any efforts to test a proposal through empirical enquiry: **how does one design empirical research to account for a pre-actual area of reality that cannot be accessed by the researcher's sensory observations?** The academic community has yet to provide a clear answer to this question. So, to progress this empirical stage of my study, I evaluated three areas of realist literature, each selected because of the efforts they make to tackle similar issues:

1 Deleuzian research methods for the social sciences
2 Critical realist research methods
3 Speculative methods

Deleuzian research methods for the social sciences

Deleuzian scholars, Williams (2008) and Bryant (2008a) hold that, whilst we cannot identify potentials in a stable environment, we are more likely to identify them following a significant change or 'shock'. This is because such shocks force entities (or assemblages) to adapt and evolve by pursuing different potentials (becomings). This argument underpins their proposals for a research strategy underpinned by a 'shock' or encounter.

In his discussion on transcendental empiricism Levi Bryant argues that we should

> seek out those gaps, events, traumas or shocks, and encounters which upset the smooth continuity of the subject.
>
> (Bryant, 2008a: 266)

Bryant's arguments are developed at a strategic level, offering researchers very few working examples to illustrate how this 'shock' strategy might be developed into an empirical research design. However, 'naturally occurring' shocks provide the basis for a number of other empirical studies. In Chapter 3 I reviewed two such studies. Bonta's study of changing land-use in the Honduras and Halsey's study of the changing Goolengook forest are both examples of studies focusing on naturally occurring 'events, traumas or shocks and encounters'.

This research strategy has been particularly influential in the social sciences. Coleman and Ringrose's edited volume of Deleuzian empirical research methods demonstrates a growing interest in the use of Deleuze's concepts amongst social scientists (Coleman and Ringrose, 2013). This collection of essays is one of the most thorough efforts to capture the latest and most advanced studies of this kind (Coleman and Ringrose, 2013). However, like Bonta and Halsey, the studies in this collection are aimed at analysing and critiquing a changing environment rather than developing tools to *engage in* that environment. Many of them also point to a growing trend for using established ethnographic practices to generate data as part of a broader Deleuzian method (Bonta, 2009). Commenting on this link, Bonta notes that

> doing ethnography is a way of letting the world wash over us, taking note of what's going on and making sense of it. . . . [Deleuze provides us with] hints at ways to figure out how the full corporeality of the world works while we're *in the world*, not just providing us with endless new ways to interpret the meanings of signs from a safe distance.
>
> (Bonta, 2009: 142 emphasis added)

This quotation suggests that existing ethnographic methods, particularly those based on participant observation provide researchers with rich, descriptive data that can then be understood using Deleuzian analytical methods. A similar point is raised by Renold and Mellor in their ethnographic study of nursery environments. Like Bonta, they suggest that ethnography's role in a Deleuzian method is to generate the richest data possible to capture the greatest range of potentials. If these Deleuzian insights are to be successful, they add, ethnographers must seek the richest data possible. This, they note, is best achieved by drawing on multiple sources of data, especially data generated from a combination of different bodily senses (Renold and Mellor, 2013).

Other social scientists have sought this richness by drawing on differing combinations of longer-established data-generation methods such as interviews, focus groups, participant observation and image making. (Blaise, 2013; Coleman and Ringrose, 2013; Dyke, 2013; Grinberg, 2013).

This link between existing methods of data generation and new Deleuzian methods of data analysis is not limited to ethnographic research designs. Cole's study, for example, is designed around an encounter between members of the Muslim community and established forms of Australian culture. The data generated in this study follows conventional practice for undertaking semi-structured interviews. Most importantly, these interviews were undertaken not exclusively for Cole's Deleuzian study but, rather, for a very different research project funded by the Australian government (Cole, 2013).

These studies raise three important methodological assumptions. First, they suggest that these existing, empirical, qualitative methods of data generation are already suited to a Deleuzian method. Yet, for many social scientists and Deleuze alike, the methods used in a research design should be drawn from the researcher's epistemological and ontological stance (Guba and Lincoln, 1994; Deleuze and Guattari, 2004b; Lincoln et al., 2011). Given that Deleuze's ontology and arguments for transcendental empiricism are constructed as a counter position to many existing forms of empirical enquiry, it is questionable why one would assume that the methods used in such enquiries could also be used in a Deleuzian method. In other words, if social scientists are to construct a Deleuzian methodology, then the methods used to generate *and* analyse data should be compatible with Deleuze's philosophy.

Second, they suggest that this suitability lies in the richness of the sensory data they generate. Whilst the use of data rich methods have proven successful in a range of social science studies, many of these studies have started with a very different epistemological stance to the one set out in Deleuze's philosophy (Hammersley and Atkinson, 2007).

Above I noted that, for Deleuze, we should seek not to understand the actual world around us but to understand how this world comes into being (becomings). Deleuze's approach to empiricism is termed 'transcendental' because he argues that these becomings lie beyond our sensory observations.[1] With this in mind, it is equally questionable why one would assume that, of all the existing methods available to researchers, those based on sensory observations would be suited to this objective. Contrary to the suggestions made by Deleuzian ethnographers like Bonta (2001; 2009) and Renold and Mellor (2013), I do not believe that adding to the richness of this sensory data can help researchers resolve this issue. Whilst much of the data presented in Coleman and Ringrose's edited volume offers a good description of the actual world identified by the researcher, it is not clear how this data would be used to identify and explore the *processes* of actualisation. This exploration of becomings is reserved almost exclusively for data analysis.

This brings me onto the third assumption. These studies suggest that sensory data contains potentials that can only be identified and explored through subsequent methods of analysis. A similar assumption can be seen in a range of empirical studies in the social sciences – particularly those applying grounded theory methods (see Charmaz, 2006). Yet, as noted earlier in this book, Deleuze holds that potentials are often hidden or lost within sensory data.

With this in mind, I believe these attempts to form a 'Deleuzian' method for the social sciences fail to capture some of the most fundamental aspects of Deleuze's epistemological position. This failure results from a focus on data analysis rather than data generation. Of course, this does not necessarily imply that all existing data generation methods are ill-suited, or that 'Deleuzian methods' are straightforward. However, it suggests that one must select methods compatible with Deleuze's position.

Critical realist research methods

Deleuze-inspired researchers are not the only ones to struggle with the difficult task of generating data based on pre-actual processes. Many social scientists working within 'critical realism', or what is termed 'transcendental realism' in the natural sciences, have tried to address this problem (Yeung, 1997). Like Deleuze's transcendental empiricism, such stances posit that the study of observable entities does not, alone, provide us with insight into the processes that formed them. And like Deleuze, they locate these processes in an area of reality that transcends empirical observation, what Deleuze termed a 'transcendental field'. As with Deleuze's realist ontology, critical realism argues that reality is not limited to one realm: the actual world we see around us. Rather, these actual forms are created from non-actual processes or 'generative mechanisms' operating in a different area of reality, which critical realists confusingly term 'the real' (Bhaskar, 1998).

According to Sayer, studies in critical realism are characterised by their search for these 'mechanisms of causality' (Sayer, 1992; 2000).

If we approach critical realism from a Deleuzian stance, it is tempting to translate these mechanisms as an equivalent to Deleuze's 'becomings'. However, if we look at these generative mechanisms in further detail, we can see why this is problematic. In his review of critical realism, Reed notes:

> Generative mechanisms operate at different levels of analysis with different 'causal powers or influences' triggered under variable contextual conditions. So, some generative mechanisms, such as models of production, class structures, labour markets, gender regimes, and surveillance systems will be highly abstract, broad ranging and structurally constraining. . . . Other generative mechanisms, such as neighbourhood networks, peer groups subcultures and household support systems, are more localized.
>
> (Reed, 2009: 436)

From a Deleuzian stance, one would not consider a neighbourhood network as *a becoming*. This is because a network is not a long-term tendency, a potential to become something, but rather a structure that influences which becomings will or will not be actualised. So, a neighbourhood network might block a potential to become diverse by resisting certain kinds of proposed housing development. Or it might encourage a potential to become safe by promoting community surveillance. Thus, if we think back to Chapter 2 and Halsey's study in Chapter 3, the critical realist's reference to a generative mechanism is much more akin to the Deleuzian concept of *the machine.*

With this in mind, Figure 7.1 shows how critical realism's 'layered' ontology might be positioned against Deleuze's three-part ontology and how the concept of generative mechanisms might be compared to the concept of machines. Using such comparisons, one might propose that the methods used by critical realists to identify machines (generative mechanisms) can be further extended into the virtual – i.e. to identify becomings (shown as a dashed arrow in Figure 7.1). However, in the following, I will try to explain why this strategy is problematic but, also, what other lessons can be learned from the comparison.

The critical realist Roy Bhaskar suggests that, to identify generative mechanisms (machines), researchers should follow a series of methodological stages. First, they should set out a 'description of significant features [within a given event]' (Bhaskar, 1998: xvii). This description is undertaken by collating 'the key components . . . the real objects of the case, for example persons, organizations and systems. They constitute structures, i.e. networks of objects, with causal powers' (Bygstad and Munkvold, 2011).

This descriptive exercise leads to the second stage, described by Bhaskar as a 'retroduction to possible causes' (Bhaskar, 1998: xvii) – or, in other words, a move from the description of an event or phenomena to an abstraction of possible causes (Yeung, 1997: 59). According to Danermark et al., such abstraction is achieved through the identification of patterns and trends in the empirical data, and the

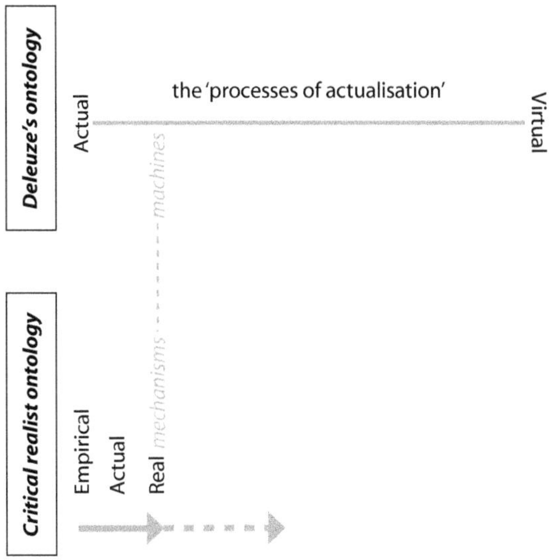

Figure 7.1 Schematic of Deleuze's and critical realism's ontological structures

linking of such patterns to existing causal explanation and concepts in the field of study (Danermark et al., 2002). According to Danermark et al., such explanations then provide the basis for what Sayer termed '*postulating* [and identifying] mechanisms which are capable of producing [the event or phenomena under study]' (Sayer, 1992: 107; Yeung, 1997: 59). In Lawson's study of housing divergence, for example, a review of the literature revealed a number of explanatory concepts that were then 'elaborated and synthesised to form a postulated causal mechanism' (Lawson, 2006: 63).

Subsequent stages of the research strategy are intended to 'eliminat[e] alternatives and identif[y] the generative mechanism or causal structure at work' (Bhaskar, 1998: xvii). For critical realists, this is made possible by collecting more empirical evidence on the phenomena or testing, revising or reaffirming the postulated mechanisms until no further contradictory evidence is obtained (Yeung, 1997: 59; Glynos and Howarth, 2007: 34).

This sequence of methodological stages highlights two ways in which critical realists respond to the methodological question I introduced earlier in this chapter – namely, *how does one design empirical research to account for a pre-actual area of reality that cannot be accessed by the researcher's sensory observations?* The first is captured in the idea that mechanisms in the pre-actual realm can be traced from empirical, sensory data either by identifying trends or by testing and refining a postulation (Figure 7.2). The second is captured by the idea of abstraction – i.e. the idea that a researcher can access the pre-actual realm by introducing abstract concepts or postulations (Figure 7.2).

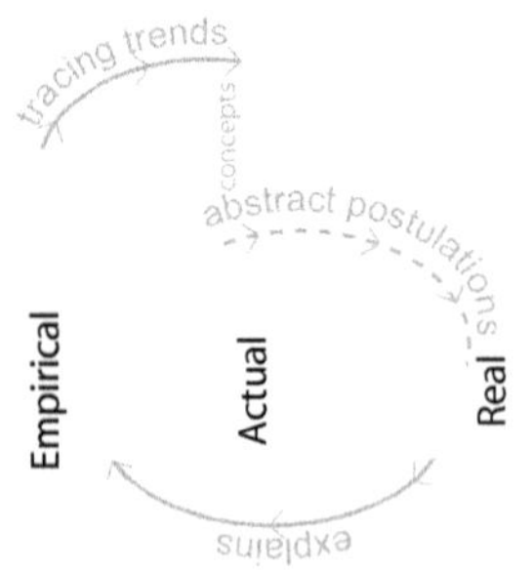

Figure 7.2 Two processes used by critical realists to access the pre-actual

For Deleuze, the first of these two methods, tracing trends, is problematic. Whilst Deleuze does not discuss critical realism directly, he does discuss the Kantian arguments on which critical theory is based. According to Deleuze, such Kantian inspired methods

> amount to both a circular argument (the conditions are supposed to account for the conditioned, yet we arrive at the condition by tracing them from the conditioned), and to arriving at the transcendental based on its *resemblance* to the actual or the condition.
>
> (Bryant, 2008b)

This circular argument within the critical realist method can be seen in Bhaskar's methodological steps discussed above and captured in Figure 7.2. For Deleuze, this circularity arises because critical realism tries to combine two incompatible ideas: (i) that mechanisms can be identified empirically and (ii) that they operate within a pre-actual, transcendental field.

Building on Felix Guattari's essay 'Concrete Machines' (1984) and his subsequent work with Guattari in *Anti-Oedipus* (2004) and *A Thousand Plateaus* (2004a), Deleuze's philosophy resolves this problem by linking these two ideas to two different kinds of mechanism (or machine): the *concrete machine* and the *abstract machine*. The machines discussed in Halsey's study are good examples of what Deleuze terms *concrete machines*. As the name suggests, such machines can be identified through empirical observation (i.e. concrete). In Halsey's case, these machines take the form of concepts identified in policy documents. If we were to locate these concrete machines in the Deleuzian ontology, we might place them close to, or even within, the actual realm (Figure 7.3).

Whilst concrete machines operate in or around the actual realm, abstract machines operate within the transcendental field (Figure 7.3). As the name suggests, abstract machines do not take on a concrete form that would allow them to be identified through empirical observation. But like concrete machines, they also

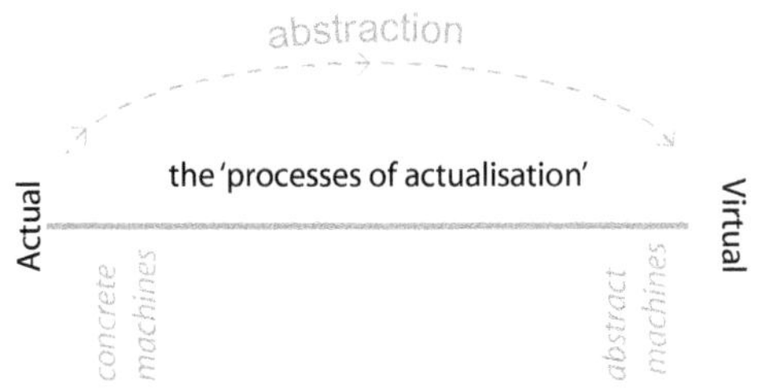

Figure 7.3 Concrete and abstract machines set within Deleuze's ontological framework

act as a kind of filter: encouraging or blocking becomings from moving across the processes of actualisation towards the actual realm (Figure 7.3). Deleuze captures this idea of the abstract machine in *A Thousand Plateaus*, when he notes:

> An abstract machine in itself is not physical or corporeal, any more than it is semiotic; it is *diagrammatic* (it knows nothing of the distinctions between the artificial and the natural either). It operates by *matter*, not by substance; by *function*, not by form. . . . The abstract machine is pure Matter-Function – a diagram independent of the forms and substances, expressions and contents it will distribute.
>
> (Deleuze and Guattari, 2004b: 156)

It is important to note that, for Deleuze, the abstract machine is very different to the referential model discussed in Chapter 2. This distinction exists because, for Deleuze, an abstract machine has no clear link to the actual entities we see around us. They are, as Bell notes, 'not to be confused with the stable forms of meaning, the established relationships between content and expression, form and substance' seen in a referential model (Bell, 2006).

For Deleuze, the abstract machines shown in Figure 7.3 and the becomings that they block or encourage can be accessed only through processes of abstraction. Deleuze's texts suggest that these processes of abstraction are dependent on the field of study. So, for example, in *What Is philosophy?* Deleuze suggests that to access the pre-actual, a philosopher must pose abstract questions (problems) and respond to these questions by creating abstract concepts (solutions). It is only by doing so that philosophers can 'lay out their own plane of immanence (their own philosophy): . . . an abstract machine of which these (concepts) are the working parts' (Deleuze and Guattari, 1994: 36).

Similarly, in *The Logic of Sensation*, Deleuze discusses the abstract process used by Francis Bacon to create his paintings. He describes this as the formation of *a diagram* (a synonym for the abstract machine)[2] through which becomings can be revealed or hidden on the canvas. The most cited of these becomings include a *becoming meat* and a *becoming animal* of the human subject (Deleuze, 2003). Whilst Deleuze uses different terms to describe these processes, this method of abstraction to reveal and select becomings for further development can also be seen in his review of Franz Kafka's literature (Deleuze and Guattari, 1986) and Luc Goddard's cinematic films (Deleuze, 1986).

The above review of critical realism and Deleuze's onto-epistemological stance demonstrates why I decided not to use critical realist methods as the basis for empirical enquiry. In doing so, it also highlights two important features in a Deleuze-compatible method of empirical enquiry:

1 out of the two critical realist methods for accessing the virtual, tracing and abstraction, only the latter is compatible with Deleuze's philosophy;
2 methods of abstraction are field dependent.

A speculative method

This latter point is important because it explains why I decided to look outside of Deleuze's texts and the references he makes to his methods of concept creation, or the artistic methods used by Bacon, Kafka or Goddard.

Looking at my own field of interest, I returned to my review of Jean Hillier's proposals for a new, Deleuze-inspired method for strategic planning. As noted in Chapter 3, Hillier argues that, to create a strategic plan, different actors should be encouraged to consider existing spatial relations and forms of practice and speculate what they might 'become' in the future – i.e. the potential for these things to operate in different ways.

In keeping with Deleuze's stance, this approach to abstraction does not trace pre-actual mechanisms (machines) or becomings from sensory observation. It does, however, differ from the methods used by Deleuze as philosopher, Bacon as artist, Kafka as novelist and Goddard as filmmaker in two important ways. First, unlike these methods of abstraction, Hillier's method is not centred on the abstractions made by a single person – i.e. the planner. Instead, this method encourages other professional actors to contribute to the process of abstraction and the becomings that are identified from that process. This distinction is important to this piece of empirical research because, like Hillier's strategic plan-making strategy, it too relies on the input of participants, albeit for the purposes of testing a proposed planning tool rather than developing a strategic plan.

The second point relates to the way Hillier envisages empirical observations and the role they play in the process of abstraction. In Deleuze's philosophical method, it is not clear how abstract questions take account of the philosopher's own empirical observations. Similarly, whilst Bacon uses live models in his work, it is not clear how the link between the empirically observed model and the 'painting as diagram' is made. Hillier's proposals are much clearer in setting out this relationship. Hillier suggests that the process of abstraction relies on the participant's memory of sensory observation. Participants then imagine the different ways in which the entities forming those observations might play out in different circumstances. In doing so, Hillier's method offers a clear use for empirical observation without falling into the circular arguments seen in critical realism. Hillier avoids this circularity by defining

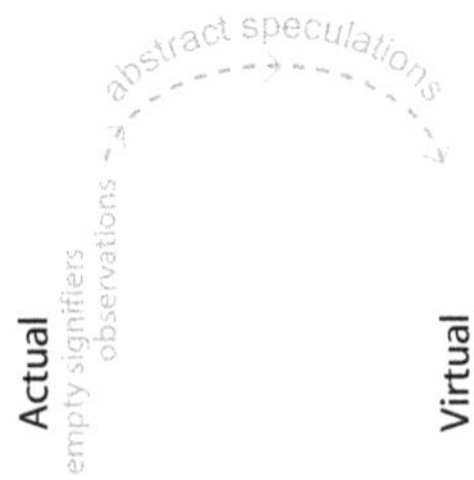

Figure 7.4 Accessing the transcendental field through speculation

a method in which speculated potentials (becomings) are *created from* observations rather than *traced from* observations (Figure 7.4).

The above figure situates Hillier's method of speculative abstraction in Deleuze's ontological framework. Unlike the other Deleuzian research methods discussed in the first body of literature, Hillier does not suggest using empirical observations as data and using becomings as the result of subsequent analysis. Rather, she seems to suggest that empirical observations provide a basis from which data (speculated potentials) is generated.

To ensure the speculative data generated serves the process of plan making, and the process of data generation, Hillier introduces a number of control mechanisms that I deemed to be equally useful to this study. As I noted in Chapter 3, Hillier argues that the process of speculating potentials 'should include concepts towards which actants desire to move such as sustainability' (Hillier, 2007: 249). Rather than looking for, or defining, an 'explicit, concise, significance' for such concepts, she argues that they should be treated as 'empty signifiers' (Hillier, 2007; Gunder and Hillier, 2009). As such, these 'empty signifiers' are able to 'secure multifarious points of view, chains of significations constituting conflicting narratives, or unique interpretations pertaining to particular situations, all under one common label' (Gunder and Hillier, 2009: 17). Methodologically, these 'empty signifiers' provide Hillier with some control over the kind of potentials generated by different actors. But it also ensures that the participants have control over the kind of potentials they speculate as relevant and the kind of experiences they use to form these speculations.

These two ideas – that speculated potentials can be taken as data, and empty signifiers can direct these speculations – were instrumental to my proposals for SIAM, as well as my empirical research. This transition between epistemology/methodology and research design is set out below.

A research design

Delphi techniques

To integrate these ideas – that speculated potentials can be taken as data, and empty signifiers can direct these speculations – one must identify a suitable research framework. I believe the Delphi techniques provide such a framework for a number of practical and theoretical reasons.

Delphi techniques are often used in studies aimed at exploring and testing new proposals through speculations (Nowack et al., 2011). Methodologists such as Wright and Rowe suggest that this suitability is owing to the structure common to all Delphi projects (Wright and Rowe, 2011; see also Linstone and Turoff, 1975; Nowack et al., 2011). All Delphi methods are formed as a collection of 'rounds'. In each round data is generated by interviewing or surveying a number of respondents, collating this data into an anonymised report and sending the report to respondents for review, thus forming the basis for a subsequent round. The anonymous nature of the Delphi rounds assures the respondents that the comments and ideas they make during an interview do not have negative consequences on their professional status. The iterative process of the Delphi rounds also means that these comments and ideas can be reviewed by other professionals working in the same or similar fields. These two techniques avoid the issues identified with interviewing as part of participatory observation and formal questionnaires because it encourages detailed discussions about specific issues in a 'live' scheme as well as broader issues drawn from multiple, historic schemes; interviews can encourage respondents to use this to identify observations, to abstract these into broader commentaries and to abstract them further to speculate about the way these points might play out in the future; respondents can omit details that would expose them to professional risk without limiting the quality of the data generated.

As well as meeting the practical demands of the research, Delphi techniques capture the theoretical demands of Deleuze's philosophy. The most striking link between Deleuze's philosophy and the Delphi techniques can be seen in the emphasis given to data as part of a creative, iterative process.

For Deleuze, problems and solutions do not exist in isolation because the relationships between them are not linear and axiomatic. Rather, he argues that they occupy different positions in his ontological structure: problems are abstract and virtual, whilst solutions are differentiated and actual (DeLanda, 2002). As noted in Chapter 2 and earlier in this chapter, Deleuze's epistemological stance refutes the possibility that the researcher, philosopher, planner or architect can fully capture the 'pure' virtuality of the problem or the 'pure' actuality of the solution. Instead, one must work with the processes of actualisation that occupy the space between these extremes. Deleuze suggests that, in order to do so, one must work creatively by identifying and experimenting with potentials from which new assemblages might emerge. But he argues that this process is not mono-directional; these new assemblages reveal previously unspeculated potentials that push us back towards the virtual problem and set out 'a new land' (Deleuze and Guattari, 2004b: 178) from which to explore other potentials and other assemblages. By moving between data generation, data analysis, further proposals for a solution and so forth, the Delphi techniques provide researchers with a method that reflects these Deleuzian arguments.

The following table shows how these Delphi techniques were incorporated into this study:

Table 7.1 Breakdown of Research Stage B into two sub-stages

<table>
<tr><td colspan="3">Research Stage B:
Is there any empirical evidence to suggest that such a tool could be made practicable?</td></tr>
<tr><td colspan="2">Research Stage B1: first Delphi round</td><td>Research Stage B2: second Delphi round</td></tr>
<tr><td>First part of the interview: Participants were asked to outline their experiences working with BRE assessments and to discuss the way such tools have affected practice and opportunities for improving these assessments.</td><td>Second part of the interview: Participants were asked to consider the fourteen statements drawn from the proposal in Research Stage A and to speculate how these statements might play out in practice.</td><td>Feedback: Each participant was asked to review the executive summary and report from Stage B1 and consider how the experiences and speculations from other participants might help them adapt, revise, expand or add to the speculations they made in Stage B1.</td></tr>
<tr><td colspan="2">Report: The principle comments and speculations from each interview were compiled into an executive summary and a comprehensive report. These documents were distributed to each of the participants.</td><td></td></tr>
</table>

Selecting participants for the study

Many key figures in the literature on Delphi techniques suggest that, to make the most of the Delphi process, participants should be selected who reflect high levels of expertise in the field (Hussler et al., 2011; Nowack et al., 2011). Criteria for selecting experts vary across Delphi methods literature. Whilst self-assessment has proven popular amongst many Delphi studies, a number of methodologists suggest that it should be avoided in favour of criteria devised to accommodate different kinds of expertise or expertise relevant to specific research problems (Hussler et al., 2011; Nowack et al., 2011). This argument for expertise is important to this study. A thorough and extensive knowledge of BRE assessments and the field in which such assessments operate is necessary if participants are to offer useful comments and speculations.

The literature on Delphi methods also acknowledges that the decision to focus on experts in the field must be balanced against their willingness or capacity to participate in the study (Bolger et al., 2011; Nowack et al., 2011). This is because experts are likely to have a high, or key, position within the industry, suggesting their time will be deemed more valuable, in terms of both the time that can be taken away from the industry and the cost of this time. This is particularly problematic in the Delphi method outlined above because it demands several rounds, each divided into different time slots.

These time constraints also affect the sample size. Even if it was possible to obtain a large sample of experts willing to participate in the first round of interviews, the use of Delphi techniques means that the results of these interviews must then be captured into a report that forms the basis for the second round. Given that such experts are unlikely to read a very long report, a balance must be struck between the number of participants and the depth of the interviews (see Nowack et al., 2011).

Taking this into consideration, I used the following criteria and method to produce the sample of respondents shown in Table 7.2 below.

Criterion 1: experience

All selected respondents have worked with one or several BRE assessments on at least five projects over the last five years. These projects include large residential and mixed-use masterplans, refurbished and new build schools, commercial and industrial schemes, and affordable and commercial housing developments ranging between one and twelve units. This criterion ensures that the comments respondents made can be generalised across their own experiences, and it reflects some of the most recent trends in the field as a whole.

Table 7.2 Experts participating in Research Stage B

Respondent	*Sampling*	*Professional role*	*Professional role*	*Completed stages*
R1	Targeted	Design and development: Strategic planner/designer	Design director in large company	Stage B1 and B2
R2	Targeted	Assessor/regulation: CSH Assessor	Founding director	Stage B1 and B2
R3	Snowball	Assessor/regulation: CSH Assessor	Founding director	Stage B1 and B2
R4	Snowball	Assessor/regulation: CSH Assessor	Senior assessor	Stage B1
R5	Snowball	Developer/client: affordable housing	Development partnerships officer	Stage B1 and B2
R6	Snowball	Developer/client: affordable housing	Senior development officer	Stage B1
R7	Targeted	Designer: urban designer	Founding director in small practice	Stage B1 and B2
R8	Targeted	Designer: architect	Architect in medium practice	Stage B1 and B2
R9	Targeted	Designer: architect	Senior director in medium practice	Stage B1 and B2
R10	Snowball	Assessor/regulator: CDM assessor	Senior assessor in medium practice	Stage B1 and B2
R11	Snowball	Assessor/regulator: planner	Senior planner in large practice	Stage B1 and B2

Criterion 2: breadth of knowledge

All respondents work in senior or project management roles. This includes several founding directors of small consultancy firms (R2, R3 and R7) and one principle director in a large national consultancy firm (R9). This ensures that the comments made by respondents take account of a broad range of concerns and the roles played by others in the project/regulation team.

Criterion 3: breadth of selection

Participants work in a range of associated fields. The sample is formed from an approximate fifty-fity split between professionals with vested interest in the design (such as architects, masterplanners and clients) and those with a role in assessing and regulating designs (such as Code for Sustainable Homes assessors, CDM coordinators and planners). This breadth reduces some of the single-party bias likely to be found in any one discipline, and it ensures that a range of interests and concerns are represented as far as is reasonable within the time constraints of the study.

Criterion 4: willingness to participate

The average duration of an interview per respondent in the first Delphi round was one hour and forty-seven minutes, and the longest interview lasted a total of three hours and thirty-seven minutes. Factoring in the amount of time needed to prepare for this interview, to read the report from this interview, to complete the resulting feedback sheet and to review the results of the second Delphi round, a conservative estimate of the average time needed from each participant is approximately three hours. Given the experience of these respondents (criterion 1), such an investment significantly limits the number of respondents willing to participate in the study.

Using a combination of targeted sampling and snowball sampling methods, I enrolled a sample formed of eleven experts across different roles in the design/assessment process, working in different sized companies or departments and with different status in those companies/departments (Table 7.2).

Notes

1 It should be noted that the use of the term 'beyond' here refers not to something outside of but to something inaccessible.

2 As noted by Deleuze himself in *The Logic of Sensation* (Deleuze, 2003) and again in his study of Foucault (Deleuze, 2006).

8 Empirical experiments

Research Stage B

Aim and structure of the chapter

This chapter presents and analyses the findings from the second research stage (B) and considers how these findings relate to the ideas presented in Deleuze's philosophy. It achieves these aims by reviewing responses to existing BRE forms of assessment and then by reviewing responses to my proposals for SIAM. Within these two sections, I have compiled, analysed and discussed the responses according to two themes:

Theme A: the relationship between sustainability assessments and the design process

Theme B: the relationship between national targets and specific contexts

These two themes are discussed in turn according to three sub-sections:

1. The identification of a problem
2. Comments and speculations about the factors causing or exacerbating the problem
3. Speculations about ways to adapt BRE assessments/SIAM to resolve the problem

In each sub-section, I bring together the initial findings from the interviews in the first Delphi round, highlight any significant points raised in the second Delphi round, and identify parts of the data that demonstrate agreement or disagreement with Deleuze's ontological framework.

The data: observations, comments and speculations about BRE assessments

Theme A: the relationship between sustainability assessments and the design process

The identification of a problem

All respondents in the first Delphi round noted that, in practice, the design process and the process of assessing a building or urban design through BRE forms of assessment are highly disconnected. Respondents saw BRE assessments as the principle source of this disconnection, describing them as 'a series of questions that you have to tick or cross . . . a very literal way of trying to extrapolate what is a very complex process' (R1); something that 'does not relate to what they (designers, contractors etc.) do in the project team' (R2); something 'impractical' (R5); something that is 'only based on stuff that you can measure' (R7); something that is 'back to front in terms of the design process' (R8); something that 'stops the building doing what it is supposed to do' (R9); a 'standalone element . . . quite isolated' (R10). Whilst using different terms, these comments suggest that this problem is evident to professionals working in the field of design and development, as well as those working in the field of assessment and regulation.

Drawing on their observations from practice, all of the respondents went on to identify instances in which design solutions were introduced to achieve assessment points rather than to meet the functional, contextual, budgetary or long-term maintenance considerations of a project. Two of these respondents suggested that some schemes were designed to meet BREEAM or CSH levels with the prior knowledge that they would be re-designed following planning approval (R7, R11). One of these respondents referred to such schemes as 'fake' master plans (R7).

The other eight respondents focused on entities introduced into a scheme to increase the assessment rating. These included the introduction of solar panels, bat boxes, bird tables, cycle hoops, bike sheds, compost bins and home offices. Drawing on their experiences and opinions, these respondents suggested that the following problems might result from such measures:

- the resulting design might have a larger overall carbon footprint;
- such measures might be used to offset very poor design decisions, especially given that assessments do not penalise bad decisions;
- such measures might clash with the architects' professional duty to the clients to meet their project brief and to achieve the best possible value for money;
- these measures might impose difficult long-term issues concerning maintenance and use;
- introducing such measures might limit the design or financial viability of achieving broader objectives, such as providing the number of homes needed to meet housing demand;
- such measures often clash with other regulations; this may cause difficulties in the design process and introduce delays to the programme/funding.

In their appraisal of these speculations, all respondents noted that assessment-driven design changes shifted the design away from, what they held to be, a 'sustainable' scheme. Respondents offered different understandings of sustainability including sustainability as holistic carbon footprint, longevity, functionality, reasonable judgement, time efficiency etc., and they noted that such definitions were not accounted for in existing BRE assessments. This range of definitions supports the argument underpinning the SIAM proposal and Deleuze's non-essentialist ontology more broadly.

Comments and speculations about the factors causing or exacerbating the problem

The responses point to three underlying issues believed to cause or exacerbate the disconnection between the design process and existing BRE assessment processes:

1. The construction of BRE assessments: some of the comments are directed at the principles used to construct BRE assessments.
2. The use of BRE assessments: the data also suggests that, in some instances, respondents see the problems resulting from use more than from construction.
3. Inherent differences in skills and knowledge: the data also points to an inherent difference between the skills and knowledge particular to assessors and designers.

These three points are discussed in greater detail below.

THE CONSTRUCTION OF BRE ASSESSMENTS

Four of the respondents interviewed in the first Delphi round suggested that the disconnection between design and assessment might be owing to the difference between a flexible design process and an inflexible assessment method. In the second Delphi round, R2, for example, comments:

> I think the main reason it (the CSH assessment) is treated as separate is due to the lack of flexibility within the assessment. A one size fits all solution does not produce good designs. You need to have a holistic view of the site, the end user and client's aspiration and attitude.

One respondent, working in a design role, also noted that most designers start by identifying a specific problem in the scheme and explore ways of resolving that problem through the design. The fixed nature of the assessment, they added, means that this individual design solution cannot be assessed without taking account of the building as a whole, which, in many instances, is not practical or appropriate (R8).

Given these limitations, two respondents describe such assessments as 'blunt tools/instruments'. R7, for example, discusses an instance in which BRE-like criteria were used to evaluate a scheme during a judicial review. After considering the evidence presented by the assessors, the inspector

> threw it out completely, and when they gave their evidence first he said, 'Look, I'm going to have to stop you there because these design criteria tools are too blunt an instrument because they don't take into account building programme. The urban criteria for a supermarket is different to a library, or a hospital'. And suddenly it was like a breath of fresh air that somebody else got it. That architecture cannot be stamped with a generic solution. Every building is different. (R7)

This experience and commentary demonstrates R7's frustration with an assessment tool that he believes is unable to capture the uniqueness and complexity of the design process. R9 expresses a similar frustration with the BREEAM criteria:

> that isn't to say that a BREEAM rated building isn't sustainable but I don't think it is optimised. You can spend money getting the points. There are other things that you could do, which in terms of sustainability would represent better value for money, but you might not get the points so it's just blunt, that's all.

In both instances, R7 and R9 suggest that sustainable assessments like BREEAM are 'blunt' because they are based on measurable traits to the exclusion of other issues in both the design and the use of the building. Such issues include the judgement made by designers in an effort to achieve 'a good compromise between (different) . . . criteria' (R9) as well as the way it will be used when it is 'up and running' and whether people will be happy using the building (R7).

These arguments against a standard 'one size fits all' assessment method found in the data captures Deleuze's stance against the referential model, articulated most strongly in *Difference and Repetition* (see Chapter 2). Like Deleuze, these respondents suggest that designs should be assessed not as 'repetitions of' a model, or judged as 'differences to' a model but as a unique response to context.

THE USE OF BRE ASSESSMENTS

The comments above suggest that the problems associated with referential models are mainly found in the assessments used to evaluate an otherwise complex and unique design (such as BRE assessments). However, contrary to R7's remark that 'architecture cannot be stamped with a generic solution', the data shows that the use of standard, referential models are as influential in the design process as they are in the assessment methods used to assess the design.

R2 notes that many large, private developer clients have developed 'standard house-types' (R2) that capture all the requirements of the BRE assessment and use these across all sites. Three of these respondents believed that these standard house-types are the results of standardised design, development and procurement processes used by this particular sector. These were described as 'a production line' (R1), a 'set' process (R2) and a 'generic' process (R7) but also as a manipulative process, particularly of the planning system (R7).

On the face of it, these critiques of 'standard designs' and 'standard processes' suggest parallels with the critique of standard models of assessment noted above. Indeed, R6 suggests that they are partly constructed in response to assessments criteria used by the BRE:

> Those regulations lead the design decisions so much of the time. We are often stood on site and we ask, does that comply with Secure By Design or Lifetime Homes? And that is the decision process. It's not about whether its right for

> the site and the end user, it's just about complying with this or that. So you are completely led. We are forever saying that aren't we?

However, the data suggests that the problems associated with standard, referential models are not derived from this link alone. As some of the comments above demonstrate, most respondents hold that the problems associated with standard designs and processes can be traced to a very specific area of the construction industry: private, commercial developments. R1, for example, notes that

> housebuilders work very much on a production line process. Their procurement is based on an end product and anything new that you bring in basically derails what is a very fixed procurement process. . . . [T]hey would just find what the cheapest way of doing it was.

Many of the respondents interviewed in the first Delphi round support the link R1 makes between the creation and use of standard models and a private, commercial company's objective to make profit. However, the data suggests that referential models are not used to design private, profit-making schemes alone. During an interview with two professionals working for a not-for-profit housing development company, it was noted that

> things like kitchens, tiles, door handles, boilers – anything that our maintenance team will need to replace on a day-to-day basis. These are standardised. This is better for us than having a different spec in every house.
>
> (R5, 6)

This quotation demonstrates that standardised interior designs and specifications are also used by not-for-profit development companies to reduce maintenance costs. Thus, whilst referential models are used by profit and not-for profit groups alike, many respondents only identify the former as problematic. The difference, it seems, revolves around an apparent distinction between 'making profit' and 'making cost-savings'.

Later in the interview, R5 and R6 offer some indications as to why these two becomings (becoming profit and becoming cost-saving) are viewed differently.

R5: We'd like to design the schemes so you can walk down the road and not distinguish tenant types. So we want our schemes to be tenure blind.

R6: . . . So a private developer will put their emphasis on profit whereas ours will be on social value and providing the things we are most proud of like sustainability. As long as it breaks even, that is our only financial interest. We don't need to make a profit on it. Our emphasis as a business is about adding value and social benefits rather than making money.

In this quotation, R6 suggests that the design of commercial schemes is problematic because the referential models used to determine the design, development and procurement process are dominated by a singular objective: to make profit, or, in Deleuzian terms, by a single *block of becoming* formed from a *becoming profit* of the scheme. In contrast, referential models used in not-for-profit schemes are conceived as a way to make cost-savings which facilitate and support other objectives, or, in Deleuzian terms, a *block of becoming* that encourages the actualisation of other becomings such as *becoming tenure blind (equal)*, *becoming sustainable* and *becoming socially responsible.*

From a Deleuzian perspective, this argument against profit confuses the ontological role played by the concept of becomings and the concept of machines. For Deleuze, a potential to make profit in itself cannot block other potentials. After all, a potential to make profit has also contributed to the actualisation of the bricks, slates and concrete used to construct the not-for-profit, social housing discussed by R5 and R6, as well as the actualisation of the consultancy and construction companies designing and building the schemes.

In Deleuze's ontology, the filtering of one potential over another is determined by machines (Halsey, 2006). In this instance, Deleuze would argue that R5's critique against private developments is a critique of several concrete machines: standard house types and standard design processes. Given that these are found in commercial and not-for-profit companies, Deleuze would argue that the same critique must apply to each. Of all the respondents participating in the research, only one respondent highlighted this issue. In the second Delphi round, R9 asks,

> Why do you single out 'large, private developers'? The public sector is at least as likely to behave cynically and manipulatively in my experience!

The data, therefore, suggests that most respondents perceive referential models in the assessment/regulatory process different from referential models in the design/development process. Problems in the former are located in the tools themselves, whilst problems lie in the dominant role played by a specific becoming. As a result the data both supports and contradicts Deleuze's ontological framework.

INHERENT DIFFERENCES IN SKILLS AND KNOWLEDGE

The third group of comments relating to the disconnection between the design process and existing assessment methods points to differences in knowledge and skills across design/development and assessment/regulatory roles.

Two respondents working in regulatory roles suggested there was a lack of technical knowledge amongst designers and small-scale developers about the assessment process. R2 recalls:

> I've sat in meetings and the client, developer or architect has said, 'You tell me what I need to do – just tell me, I don't want to have to choose. You choose – tell me what is the simplest, quickest, cheapest way to get Code level 3'. . . . The main problems with Code is the small developers in our experience because they don't know about it, they don't understand it.

However, they also acknowledged that this knowledge deficit might result from the language used in BRE assessments. Such language, they argued, is highly technical and does not reflect the language used by those working in design, development and construction roles. Using a simple analogy, R2 discusses the difference between the language used in the BRE assessments and the language used by designers, developers and user groups:

> I suppose the problem is that there are two people using different kinds of language. Going back to the jeans/Gap example, the first person [representing the designer or user group] is asking, 'Are these comfortable?' But the shop assist [representing the CSH assessor] is saying, 'Well these jeans have a low crotch and a wide boot cut'. And the first person will think, 'Does that mean that they are comfy or not?' They are using different language – 'I use these words, are they the same?' They don't quite match up.

This analogy suggests that assessments do not correspond to the immediate, practical issues facing the design of a building. This same problem is expressed differently by two other respondents working as design professions. They argue that the disconnection between design and assessment results from a lack of awareness about how assessments affect a given design. For one of these respondents, this is particularly problematic because most designs are assessed by 'someone who doesn't understand the [design] decision making process' (R7). Whilst for the second respondent, the problem lies in the lack of design awareness applied to the formation of a company's environmental policies. This, they added, is limited not only to the policies used by clients to set-out a design brief but also to the policies used by designers and developers in order to determine how they will respond to a design brief (R9).

Whilst approached from a different stance, all four of these respondents suggested that to overcome the deficit in knowledge and, thus, help resolve the problem of disconnection between design and assessment, the BRE should change their assessment methods to better reflect the design, development and construction processes.

For several respondents, the best way to do this is by re-thinking the role played by assessors. These respondents argued that problems with BRE assessments were exacerbated by what they perceived as the CSH or/and BREEAM assessors detachment from the design process. Because the BRE assessors work 'outside of the design team' (R10), two respondents in particular felt that the assessors do not have a nuanced understanding about the process through which design decisions are made. As a result, they held that assessors are unable to make informed judgements about the relevance of assessment criteria. R10 expressed this position clearly when he notes:

> [The] CSH assessor seems to be, pretty much, a stand-alone element – not even of the design team. They seem to be quite isolated. Whilst they provide comments and information to the overall project they are not as coordinated

> within the team as everyone else in the design team. . . . They chip away from the outside but don't actually understand what needs to be done and the issues in the scheme.

Drawing on their experience as BRE assessors, two respondents argued that the assessor's detachment from the project team was not intentional. Rather, it was the result of two factors: the construction of the assessment, which favours completed buildings over work in progress; and reluctance amongst members of the project team to engage in discussions that they feel do not attend to the immediate issues facing the design, development and construction of the scheme.

> The best people to speak to about this are site managers – I went to visit a site in London and the site manager said, 'All this sustainability stuff, do you know how much I need to think about and do, and all these charts I have to work to? I have the welfare, and CCS man coming – I have guys out there all need managing and now you are coming here saying I need to do this that and the other'. They hate it. It's just more things to think about while they are building.
>
> (R2)

Speculations about ways to adapt BRE assessments to resolve the problem

Developing these observations, comments and speculations further, respondents considered how the existing BRE suite of assessments might be re-directed to help resolve this problem of disconnection between assessment and design. These recommendations fall into two broad categories:

1 Standardising and simplifying assessments
2 Adapting the method to allow for different contexts and roles

STANDARDISING AND SIMPLIFYING ASSESSMENTS

Three respondents suggested that simplifying the BRE assessments by focusing on CO_2 alone might make the assessment easier to understand and to 'roll out' across all design/development/construction and regulatory industries (R1, 2, 3).

R1 stressed this argument throughout both Delphi rounds, claiming that a focus on CO_2 would capture many other design objectives:

R1: . . . So, if you create a place of good design quality, a sustainable place to be and to live that and gives you a sense of wellbeing when you are there, and all of that is embodied in its sustainable value. And I wouldn't necessarily say that I like living in that place because it has good carbon value, but its carbon value, hopefully, if you get all the ingredients right, it would also be a sustainable place to live and therefore it would have a smaller carbon footprint. . . .

GA: But how do you justify those links or translate between a sense of place or a sense of community, or whatever, which are abstract ideas and a carbon level? It doesn't seem to translate easily.

R1: It does though. If you think of a green strategy for a place, for example, now we all know that the healthier people are and the more leisure pursuits they engage in – the happier they are. If they have really high quality green spaces where they live the chances are, they are going to be happier. And the greener the place is, the smaller its carbon footprint. Because you are providing the green spaces, the trees etc. that will have environmental value and so I think the two things are inextricably linked.

This short exchange highlights an underlying and questionable argument within R1's suggestion for a carbon-focused assessment method. R1 sets out a set of relations in which a low-carbon design inevitably leads to a sustainable design; an environmentally conscious design; a design that increases happiness, wellbeing and health. In doing so, R1 points to an inherent set of relations not dissimilar to the essentialist models used to explain 'a sustainable home' (Figure 2.1) or a 'sustainable building' (Figure 2.2). Therefore, it seems that, for R1, some form of essentialist model is an inherent feature of effective assessment design.

The idea that assessments should be focused on CO_2 draws an interesting parallel with the perceived differences between profit and cost-savings. Like profit, CO_2 is offered as a dominant objective for assessment methods. But unlike profit, this objective is seen as a way to facilitate other objectives. R1's comment suggests that this is possible because of inherent (essentialist) relationships between CO_2, health, happiness, sustainability and wellbeing.

In the feedback to the second Delphi round, R11 notes:

> From a planning perspective there are three dimensions to sustainable: economic, social and environmental. If an alternative assessment method is to be devised (i.e. SIAM) it should include a variety of indicators to ensure that national policy targets can be balanced against the social, economic and environmental benefits of a development project.

In this response, R11 draws on the normative, three-part definition of sustainability used in the UK's National Planning Policy Framework (DCLG, 2012b) but also found in wider literature (see OECD, 2009 for example). This suggests that the modelled concepts used in BRE assessments are part of a larger 'network' of modelled concept definitions related to different scales, roles and uses of the sustainability concept.

ADAPTING THE METHOD TO ALLOW FOR DIFFERENT CONTEXTS AND ROLES

The data also shows a great deal of support for a very different method, one that challenges the kind of essentialist principles advocated by R1 in particular. Rather than focusing on one objective, many respondents outlined their vision of a sustainable assessment that could adapt to different professions, contexts and judgements.

Three respondents suggest that BRE should tailor the method and the understanding of 'sustainability' to different professionals at different parts of the design process (R2, 3, 10). R2 notes:

> If everybody is your customer then nobody is. It's (CSH and BREEAM) trying to be one size fits all sustainability – and you can't do that. It's different for different people, not just that but it does not relate to what they do in the project team. If you speak to contractors they start to speak about materials 'well I've got this wood and it's FSC registered. So does that mean I get CSH certified?' But the architect says 'I've got 100mm mineral wool in the wall so does that mean its Code level 3?' The site manager says, 'Well, we cover the tips up and all our boys come from the local area', because that is what sustainability is to them. So everyone has a different understanding about what it is. But then, we [as assessors] come along and say, 'No. You've got sustainability all wrong. This is it'.

In this quotation, R2 uses personal experiences working with a range of design and construction professionals to argue one of the key points within the SIAM proposal and within Deleuze's ontology more broadly. R2 argues that concepts like sustainability have no inherent meaning. Not only do they mean different things to different people, but R2 also suggests that the concept of 'sustainability' cannot be divorced from the decisions those people make *during* a design or development process. This idea is elaborated further by R2, who goes on to suggest that BRE assessments would work much better if they could be used early and throughout the design.

Building on the idea of responsibility, R9 suggested that BRE assessments would be improved if they took account of the 'reasonable judgements' (R9) made during the design process rather than focusing on the results of those judgements – i.e. the building or urban design:

> All you are trying to do really is demonstrate that what you did was reasonable. You'll never make all the right decisions in any job. And all you are looking for really is, first of all a methodology that assists you in making as good a decision as you can. And secondly, you are looking for something that leaves a record that that is what you have done.

This idea of 'reasonable judgement' is emphasised further in the second Delphi round. R4, for example, suggests:

> The assessment needs to give some autonomy and ownership to the assessor and designer. They should be able to make their own minds up whether a design has achieved certain criteria and allowing flexibility if for some justifiable reason it cannot achieve certain criteria.

Developing this position, four respondents went on to suggest that a more sophisticated approach to sustainability assessment might favour some form of self-regulation. Of these, two suggested that this could be underpinned by registration to an accredited body (R2, 3), and two suggested the creation of a new scheme focused on the design and assessment processes (R9, 10). This last stance is articulated most clearly by R9, who notes that

> a different way that you might consider this is rather than comparing it with CDM, to compare it with ISO 9001.[1] In ISO 9001 you establish a system: a quality system. The quality system doesn't mean that every decision you make is always right. But it does say that these are the standards in terms of the process that we are setting for ourselves. And this is the standard that we have got to always achieve. That is a standard determined process – 'we will always follow this process and the decisions we will make, are made as a consequence of this process' – but that doesn't mean that all decisions will be right.

For R9, the idea of reasonable judgement and self-regulation are already captured in a previous form of BRE assessment introduced in 2002 as the BRE Checklist model and superseded by subsequent developments of BREEAM (R9).

Whilst R9 provides a strong argument against essentialist modelled concept-based assessments, his subsequent calls for quantitative measurement and criteria suggest a return to these principles:

> I think that the BRE Checklist is of great value and I do not understand why it is not used more as the comprehensive framework for assessing sustainability. I would like to see it expanded and developed to include more objective methods of quantifying or scoring each criterion.

Theme B: the relationship between national targets and specific contexts

The identification of a problem

The second theme emerging from the data relates to two scales of policy/plan-making: national and local/context specific. Four respondents participating in the first Delphi round discussed these two scales in detail. They describe national sustainability targets as ways to 'focus the mind' (R1), to 'raise the bar [in sustainable design]' (R1), to ensure that 'things get done' (R2), to avoid extreme differences between projects (R8) and to demonstrate that developments are 'driving in the right direction' (R10). Yet, in developing these comments further, three of these respondents suggest that these targets do not translate to local, contextual considerations. They note that 'strategic targets can lead developers to chase the wrong goal – not design to the end user' (R1). Targets can also be 'restricting in some instances' because developers only work up to the target rarely beyond the target (R2).

The data suggests, therefore, that there is a broader problem affecting BRE assessments, concerning the link between national and local scales, and the way this link plays out over time. R1 describes this as a 'fight between national policy and local groups' and, later, as a gap between both scales of policy/plan-making.

Comments and speculations about the factors causing or exacerbating the problem

Respondents identified three factors they believe might be causing or exacerbating the conflicted relationship between national and local policy/plan-making. As in the last theme, these can be summarised according to two underlying themes:

1 The construction of BRE assessments
2 The use of BRE assessments

THE CONSTRUCTION OF BRE ASSESSMENTS

R10 suggested that the problem between national and local scales might be owing to an underlying assumption in many assessments, including those set out by the BRE: that the tools used to assess and engage in a specific context can be translated to broader national targets and vice versa.

R10:	It's probably better to think of those (national targets) as the key drivers, the large targets for the HSE is to reduce deaths on construction sites and in maintenance.
Interviewer (GA):	But, given your earlier comments, the way it's operationalised (CDM) seems much more developed – more complex and sophisticated.
R10:	Yes, much broader brushed. Things such as repetitive strain disorders and sunlight disorders are all considered as part of our role.
GA:	But how would you link sunlight disorders to the number of windows in an office building? And how would you link that to death or injury? It's difficult if not impossible. So, for me, there must be something else going on there in the operationalising of the CDM that is more than the target set by government.
R10:	Yes, true. But that is the guidance, it's about being part of an organisation whose objective is to improve H+S in the work place, in social environments. So, it's not a set of rules in the guidance.

This respondent goes on to suggest that this assumption is misplaced and that national targets and contextual assessment tools should focus on different levels of detail used in different ways.

THE USE OF BRE ASSESSMENTS

For others, like R1 and R11, the problem relates not to the principles used to construct assessments but to the way they are used by specific groups. For R1, the problem between national and local scales is exacerbated by the normative practices used in the determination of a planning application. Whilst several key planning documents, such as the National Planning Policy Framework (DCLG, 2012b), stipulate that proposals should 'defer to local knowledge', R1 noted that planning inspectors still prioritise national targets over local concerns.

Looking at the problem from a slightly different angle, R11 suggested that the problem might be exacerbated by the kind of groups driving developments towards national targets. According to R11, BRE benchmarking, as a driver for national sustainability targets, is mostly used as a condition of development set out by the developer (by the environmental policies defined by large commercial companies and housing associations). They are rarely set out as a condition attached to a specific site of development. As such, these targets are driven by national rather than contextual actors.

Thus, as with the disconnection between design and assessment, these comments suggest that the difficult relationship between national and local scales can be traced to issues associated with the construction of assessments as well as the way these assessments are used by certain groups and the priority they give to certain objectives (or becomings).

Speculations about ways to adapt BRE assessments to resolve the problem

The data suggests four ways in which the BRE might develop their suite of assessments to help resolve this problem. Many of these suggestions reiterate the suggestions made for resolving the disconnection between design and assessment. These suggestions are as follows:

R2 suggested that targets should have a motivational role in sustainability assessments, and considered whether it would be better to introduce abstract or purposefully unachievable targets. This proposal corresponds with one of the ideas in SIAM – namely, using the concept of 'sustainability' as an empty signifier that assessments are directed towards but that is too vague to be achieved in a definable way.

R10 suggested that the BRE re-think the link between national targets and assessments in order to produce a more sophisticated assessment tool, and advocated the CDM form of assessment as a model for future development.

R9 notes that 'BREEAM falls down because you are trying to get over a threshold'. As in the last theme, this respondent suggest that the BRE suite of assessments might be improved if they were to return to the method used in the BRE Sustainability Checklist, which does not use benchmarks.

Synthesis of both themes relating to existing BRE assessments

This first section of this analysis has highlighted the problems associated with BRE assessments and sets out the observations, comments and speculations made by the respondents to explain why these problems occur and how they might be resolved. The following five statements capture the key points raised in the data.

1 The use of referential models in assessment methodology is problematic because they do not allow for the different potentials captured in the design process.
2 Referential models are only problematic in the design process when they are created and used by private developers. This is because they are developed according to one kind of potential, the potential to make profit.
3 Referential models can be beneficial when used by non-profit developers, because they facilitate other potentials.
4 Assessment methods can be improved by changing the referential model so that it is focused on CO_2 or broader definitions of sustainability.
5 Assessment methods can be improved by removing fixed criteria (the referential model) and adapting the methodology to different contexts and roles. Such improvements would favour 'reasonable judgement' and self-regulation.

These five statements reveal inconsistencies within the data concerning the existing and potential role played by referential models and, more broadly, by the role played by essentialism. Of the five statements, only statements 1 and 5 capture Deleuze's ontological stance and my proposals for SIAM. Statements 2 and 3 reflect views which confuse or blur the roles played by the ontological concepts of becomings and machines. The source of this confusion is not clear, nor is it clear why this position differs from the arguments raised against assessments in statement 1. It is possible that this confusion and contradiction within the data may stem from more subtle professional bias or/and perceptions within certain areas of the construction industry. This finding would need to be explored further in subsequent research.

Of the five statements, statement 4 is the least compatible with Deleuze's philosophy. It reinforces the essentialist position underpinning existing BRE assessments, suggesting that an attempt to better capture inherent traits would result in a better model, and thus a better method.

The data: observations, comments and speculations about SIAM

This third section of the chapter turns to my proposals for an alternative method used to assess the sustainability of a building or urban design. This proposal is termed 'the Speculative and Immanent Assessment Method', or SIAM, as developed using Deleuze's ontological concepts (see Chapter 6).

Theme A: the relationship between sustainability assessments and the design process

How might SIAM resolve problems identified in BRE assessments, and what new problems might be introduced by SIAM?

All respondents felt that the philosophy and construction of SIAM better reflected the design process than existing BRE assessment methods. Respondents referred to SIAM as a 'design led' assessment with a 'more fluid approach than CSH and BREEAM' (R1); as 'integrated into the design before building starts on site' (R2); as 'a common sense assessment. There is no common sense in the Code for Sustainable Homes' (R2); as 'a really good concept – a toolkit . . . [that] makes decisions with sustainability in mind' (R5); as an assessment 'rooted in architectural thinking itself' (R7); as an assessment that 'allows you to address real situations' (R8) as well as the architectural process, particularly in respect to the 'the speculative aspect . . . because the building doesn't exist yet' (R8); as 'a more sophisticated approach [to sustainability]' (R9); as a 'more rounded approach [than BRE assessments]' (R10); as an assessment that 'is all about balance . . . taking a balanced view (on) . . . the most appropriate scheme' (R11).

Comments and speculations about the benefits and problems with SIAM

Responsibility and ownership was cited by many respondents as one of the factors in SIAM that might usefully ameliorate the problem of disconnection between design and assessment seen in existing BRE assessments.

All respondents made reference to responsibility in the first Delphi round of interviews, and seven respondents discuss this explicitly. One such respondent (R10) notes that

> SIAM is about making people responsible for their own decisions and actions . . . a sustainability responsibility.

Several respondents extend this idea of responsibility to consider subsequent impacts on the design profession as a whole. R7 noted that the architect's role has diminished over the last few years, and SIAM could help re-valorise the profession:

> Maybe you need to give more control to the architects – rather than a profession falling apart; maybe it is consolidated by something around being more responsible for everything that goes on.

Whilst the data shows overwhelming support for the ideas used to construct SIAM, it also highlights a number of concerns relating to its use in practice. Most respondents highlighted the difficulties associated with the introduction of SIAM into existing design, development, construction and regulatory practices. One such respondent (R7) noted,

> I think it [SIAM] is fantastic, but it falls down when you have to plug it into the system. The current system can't cope.

The difficulties associated with the introduction of SIAM into practice vary across respondents. Five respondents believe that SIAM's open and self-regulatory approach might be manipulated by a number of different actors, including large, private developers (R2, 3, 5, 6), designers (R2, 3) and the appointed sustainability coordinator (R1, 2, 3). However, these respondents also suggested that introducing a robust framework for self-regulation, and targeting the right audience for SIAM, might overcome these risks.

R1, 5 and 6 argue that SIAM might be incompatible with existing methods used to implement and record progress against broader government policy, whilst R2 and 3 expressed their concerns that actors in design, development, construction and regulatory roles might not have sufficient imagination to assess sustainability on a case-by case basis – i.e. without a universally applicable description and corresponding set of criteria:

> If there wasn't one model then I'm not sure people and society are sophisticated enough to come up with their own way to improve sustainability. So it probably isn't perfect but the Code is best for the most number of scenarios. (R3)

Another area of concern related to the time pressures needed to complete the SIAM assessments, particularly on simple schemes and in the early design stages of larger schemes. However, all four respondents who expressed these concerns (R2, 8, 9, 11) went on to suggest that if practices *were* willing to dedicate time to SIAM, these assessments could reduce the risks associated with planning approval, with late and costly changes to the design and with providing an audit trail for decision making if disputes were taken to court.

The data reveals instances in which similar, and equally robust, methods of assessment are already used in practice, albeit on a case-by-case basis. Four respondents noted similarities between proposed SIAM assessments and the reports used to request a waiver of conditions attached to a planning approval (R5, 6, 8, 11). All of these respondents suggested that such reports were very helpful in the relationship between design, development and regulation. Drawing on experiences as a designer, R8 suggests that this is because

> if you can understand what is going on, it's explained, and it's real then you can accept difficult things.

From a regulatory perspective, R11 felt that such reports 'tell the story' of the scheme, which, unfortunately, is an aspect that is often overlooked in the current design and regulation process. For R11, this is particularly evident in the use of Design and Access statements submitted as part of a planning application to provide evidence that the proposed scheme meets current planning policy and regulatory requirements:

> The D+A statement is usually just 'here is the site, here is a scheme, it uses a similar palette of materials to that house over there, landscape is beautiful, it's a great scheme, it uses modern architecture, please sign up to this now'. They miss that key storytelling stage of, actually, how you came to this scheme on the site.

Speculations about ways to improve SIAM

These issues led several respondents to speculate opportunities to improve SIAM as a practicable assessment tool. The four most developed suggestions are set out below.

THE ROLE OF THE SUSTAINABLE COORDINATOR

Five respondents suggested that the sustainability coordinator role should be expanded to make them more accountable. Just over half of these respondents suggested that the sustainability coordinator should 'sign off' the final building design (R5, 6, 7), whilst the remaining respondents argued that they should 'sign off' the design process (R9, 10). Of these two groups, the latter is a better reflection of the SIAM proposal.

NEW DIRECTIONS FOR SIAM

Rather than an assessment tool, respondents suggested other directions for operationalising the ideas underpinning SIAM. Respondents outlined three such directions. The first suggestion builds on speculations directed at BRE assessments in the last section of the report.

R9 and R10 suggest using SIAM as a foundation for a new ISO registration, possibly linked to ISO 14006: 2011: *Implementing and managing eco design* (ISO, 2011). These two respondents differed in the way they conceived this ISO. R9 suggests that a SIAM-informed ISO would replace or, at least, offer an alternative route to assessment:

> a new ISO for sustainable design would be a valuable outcome of this type of research . . . [with] more safeguards in terms of the overall approach to the sustainability of the building than you would if you used a point scoring exercise.

Whilst, for R10, this ISO system would be used to complement a SIAM-like assessment method:

> SIAM is the model that the coordinator uses to assess the sustainability of each project and that focuses the ISO audit to make sure that the guidance from government is being applied.

This comment from R10 seems to suggest that SIAM could be situated within a broader essentialist framework. As noted in previous areas of the data, this seems to suggest reluctance to entirely abandon essentialist, structuring principles.

Rather than using SIAM to set out a new assessment method, another suggestion outlined in the data focuses on possible links with the existing BRE suite. Three respondents suggested that SIAM might have a greater impact if it was used to change existing BRE assessments rather than provide a new, stand-alone assessment (R5, 6, 9). This point is captured in the following extract taken from an interview with R5 and R6:

R5: *The problem is how you get those people on board – BRE and those using these measures.*
GA: *So you think I need to find a way for CSH and SIAM to link in some way?*
R5: *To take the good bits from both and combine them.*
R6: *Yes – find a way for them to come together . . .*

Thus, whilst many respondents believed that SIAM could be developed in new directions, the two directions outlined in the data point to very different kinds of method. One pushes SIAM further away from the essentialist model used in BRE assessments, whilst the other looks to combine SIAM into the BRE model.

A BROADER SCOPE FOR SIAM

The third suggestion points to a very different role for SIAM. For R7, SIAM might be best used as a new method for teaching potential architects and urban designers about the design process. This teaching method could, he suggests, help students understand how design practitioners balance different interests.

Several respondents also suggested that SIAM might be more successful if it expanded the focus of assessment. Respondents offered three ways in which this might be achieved.

R7 discussed a growing trend in metropolitan planning authorities to draw on expertise in non-spatial disciplines, such as those working in the field of health and wellbeing. SIAM, it was speculated, should not focus on

> just one thing but several things. And that is the mistake we are making at the moment, we are just thinking about one thing at a time.

In Deleuzian terms, this quotation suggests that the problem with current assessments is that they are focused on one single block of becoming. A similar point is made by two other respondents who suggest that SIAM could help merge or link existing roles. One of these respondents suggested combining the role played by the sustainability coordinator with the CDM coordinator (R10), whilst the other suggests using SIAM to form the Design and Access statements used by planning consultants (R11).

TARGET MARKET

An important point raised in the interviews concerns the audience for SIAM assessments. For many respondents, the success or failure of SIAM would depend on its target market. Most respondents identified markets best suited to a speculative and immanent approach to assessment. Such markets include one-off house builders (R2, 3), small-scale developers (R2, 3, 5, 6), affordable-housing groups (R5, 6), government-led projects (R7, 9), renovations (R7), home improvements

(R8) and base builders (R8). Whilst these markets differ, all respondents identified these markets as areas in which bespoke designs are deemed important or necessary, and all felt that the existing BRE assessment method could not accommodate the bespoke nature of such markets.

Theme B: the relationship between national targets and specific contexts

How might SIAM resolve problems identified in BRE assessments, and what new problems might be introduced by SIAM?

As in the review of BRE assessments, there was a mixed response to the way SIAM responded to the relationship between national targets and a specific context.

Three respondents felt that SIAM did not provide suitable links between a specific scheme and national targets (R1, 2, 3). As a result, these respondents suggested that SIAM might be problematic for policymakers and planning departments, who could not use SIAM to demonstrate that developments were making progress as part of a national sustainability strategy:

> How do they [planners] weigh up if it's acceptable or not. Because at the minute they can refer to policy and say, 'yes it meets policy' or 'no it doesn't', whereas if they were having this evidence presented to them, would there be any consistency amongst planning officers? What would they be assessing against to say whether or not it is sustainable?
>
> (R6)

A different line of speculation is offered by R11, who suggested that SIAM could 'assist with knowledge sharing and an education piece for the whole of the project's lifecycle' (R11).

Another area of concern related to the use of national targets to direct funding to housing associations (R5, 6, 8). These respondents felt that removing the link between assessments and targets might be problematic for funders but not necessarily for housing associations, who might welcome a funding framework favouring design decisions (R5, 6).

Rather than seeing this lack of comparability as an issue, R9 and R10 argue that clear links between national targets and context-specific schemes are not necessary as long as a scheme could provide robust, demonstrable evidence that all reasonable efforts have been made to introduce sustainability into the decision process (R9, 10). For these respondents, SIAM could be used to provide such evidence.

> You need to be able to demonstrate that you have taken things into account and you've made the best, no, not even the best decision, you've made a justifiable decision. . . . You could use this [SIAM] to demonstrate that you've done your best.
>
> (R9)

Speculations about ways to improve SIAM

Drawing on these comments and speculations, respondents identified three opportunities to improve SIAM as a practicable assessment tool. These are outlined below.

INTRODUCING QUANTIFIABLE MEASUREMENT

Three respondents suggested introducing some form of measurement into the assessment (R1, 5, 6). This suggestion gained greater support in the second Delphi round, in which R2 makes the following comment:

> I think the idea of introducing some form of quantitative measurement would be a good idea.

R3 also offered some support to this proposal:

> Yes I agree that to take it forward, it would need to be adapted to create links or introduce a quantitative form of measurement [to avoid manipulation].

AN ADVISORY LIST

Rather than a fixed set of quantifiable criteria, R2 and R3 suggested introducing an advisory list, the contents of which could broadly relate to national targets. In developing this idea further, it was suggested that items could be added or removed from this list to suit changes in the design.

A ROBUST PROCESS FOR DECISION MAKING

R9 and R10's suggestions for SIAM provide a very different direction for development to the ones noted above. Building on their proposals for self-regulation and ISO approval, they suggest evaluating the robustness of the design process used rather than the resulting building. Such processes, it was argued, could be very 'broad brush' or/and unique to different companies working on given schemes (R9, 10).

As in previous themes relating to SIAM and to BRE assessments, these suggestions reflect support for the essentialist models used in existing BRE assessments, and arguments for a shift away from such models.

Synthesis of both themes relating to SIAM

This second section of the results and analysis chapter has highlighted the benefits and problems associated with my proposals for a Speculative and Immanent Assessment Method and sets out the observations, comments and speculations made by the respondents to explain why these problems occur and how they might be resolved. The following six statements capture the key points raised in the data:

1. SIAM is much closer to the design process than the method used in BRE assessments.
2. SIAM places responsibility with individuals.
3. SIAM resolves many of the issues raised against BRE assessments.
4. Whilst SIAM is a strong proposal, it is difficult to operationalise; part of this difficulty relates to the risk of manipulation and its incompatibility with existing government targets.
5. SIAM might be improved by introducing quantitative measurements and by focusing on the traits used in the assessment.
6. SIAM might be improved by focusing on the robustness of the design process and the decisions taken rather than the outcomes of that process – i.e. the building.

Whilst the first four statements support SIAM and the Deleuzian concepts that underpin SIAM, the last two statements point to very different stances: one consistent with Deleuze's ontology and one in conflict with Deleuze's ontology. These two proposals for SIAM reinforce what seems to be an underlying tension in the responses. This tension centres on the role played by essentialism.

For some, the essentialist modelled concept seems to be integral to the way they understand and engage in professional practice. This is because such models allow for comparison between schemes as well as comparison against targets and ideals of best practice. For others, such comparisons are less important than designing a scheme that maximises the potentials of the site, user demands, health and safety and budgetary limitations.

Interestingly, these two stances (essentialist and non-essentialist) do not correspond to a professional role. The data shows that professionals working in a regulatory/assessment capacity have been as critical of essentialist models as those working in a design/development capacity. Equally, some of the respondents working as designers and developers have been the strongest advocates of essentialist principles.

Conclusion to Research Stage B

Taken together, the two sections that form Research Stage B provide some evidence to support the possibility that the Deleuze-inspired proposals created in Research Stage A (Chapter 6) can be translated into a practicable tool useful to actors in the planning and development process. However, it also suggests that making this transition to practice will demand a number of changes.

One group of changes proposed in the data is more-or-less in keeping with the Deleuzian concepts used to create SIAM. These include efforts to link with, or draw lessons from, similar or complementary frameworks (BRE Sustainability Checklist, CDM assessments and ISO ratings), as well as select enhancements that take account of its use in specific areas of practice (using SIAM to inform other practicable tools, broadening the remit of SIAM, targeting the right markets, developing complementary software). Of these suggestions, I find the parallel to the BRE Sustainability Checklist to be particularly important to the line of enquiry underpinning this empirical stage of the research – namely, whether SIAM can be operationalised into a practicable tool. There is nothing in the literature to suggest why the BRE decided to pursue the essentialist 'universal' method used in BREEAM over the less/non-essentialist method used in the Checklist. Understanding this decision might provide further clues as to whether or not an assessment method like SIAM could be operationalised. This could be a subject for further research.

The second group of changes proposed in the data contradicts many of the non-essentialist concepts structuring Deleuze's ontology and used to create SIAM. These include the search for, and subsequent use of, a universal definition of a 'sustainable design' (possibly based on CO_2) and the use of quantitative information related to the properties of the proposed building or urban design. Such changes would bring SIAM closer to BRE assessments. These speculations within the data suggest that Deleuze's ontological concepts can be made useful to planning practice, but doing so may demand significant compromises to the ontological arguments underpinning the creation of these concepts. These themes will be discussed further in the following chapter.

Note

1 ISO 9001 are standards that set out the requirements of a quality-management system.

Part 4

Synthesis, discussion and conclusions

Part 3 presented and analysed the results of a case study. In this fourth part of the book, I bring these results together, discuss them in the context of the two research questions introduced in the introduction, and expand these results to reflect on some of the broader issues raised across all the chapters so far. I will use these discussions to reflect on the lessons learned from this study into Deleuze and planning before outlining limitations and directions for further research.

9 Synthesis and discussion

Introduction

In this chapter I synthesise the findings from Part 3 and my review of other Deleuze-inspired theorists in Part 2 to answer the two research questions that introduced this book.

Synthesis: answering the two research questions

In the introductory chapter I set out two overarching research questions that I have expanded and developed over the last eight chapters:

1 Can Deleuze's philosophical concepts be translated into a new theoretical framework for constructing formal assessments?
2 If so, can this be translated into a practical tool useful to actors in the planning and development process?

To answer these questions, I start by reflecting on the specific case explored in Part 3 and expand this line of enquiry to consider other forms of essentialist and non-essentialist assessments used in practice.

A theoretical framework for sustainable assessments

My proposal for a Speculative and Immanent Assessment Method (SIAM) in Chapter 6 demonstrates that, at a conceptual level at least, Deleuze's philosophical concepts *can* be translated into a new theoretical framework for constructing a formal assessment tool used to assess the sustainability of a building or urban design. The degree to which this framework was or was not successful was determined through empirical testing in the subsequent Research Stage B discussed in sub-section 'A practical and useful tool' below. The development of this first, conceptual proposal reveals three strategies used in this translation process:

Strategy 1: recreating Deleuze's concepts
Strategy 2: borrowing Deleuze-inspired concepts from other studies
Strategy 3: learning from other tools used in the field

Strategy 1: recreating Deleuze's concepts

The first of these strategies was discussed in my review of three Deleuze-inspired spatial studies in Chapter 3. This review showed that all three of the reviewed studies produced their respective theoretical framework by interpreting and adapting Deleuze's concepts to meet the specific demands of their enquiry. Whilst all three studies draw on Deleuze's concepts of the 'assemblage' and 'becomings', for example, their definitions for these terms are dependent on their role in the respective study.

In Chapter 6 I also adapted these concepts to discuss and develop an alternative to current BRE assessment methods, notably in subsections 'A practical, useful sustainable assessment tool' and 'A practical, useful assessment tool more broadly'. In these two subsections, I drew on Deleuze's conceptual distinctions between 'objects' and 'assemblages', 'properties' and 'becomings', and 'referential models' and 'blocks of becoming' to set out a comparative synopsis of the BRE form of assessment, and my proposals for an assessment method constructed using Deleuze's concepts: SIAM.

Strategy 2: borrowing Deleuze-inspired concepts

The second strategy I used to make this transition relied on concepts borrowed from other Deleuze-inspired studies and adapted to suit my own framework.

In Chapter 3, I identified concepts in all three frameworks that were created specifically to resolve conceptual or practical limitations identified in Deleuze's texts. In my review of Mark Bonta's study (2001; Bonta and Protevi, 2004), I showed how Bonta expanded Deleuze's concept of assemblages to form the concept of 'complex space'. For Bonta, this concept better reflected the process of entanglement identified in his observation of land use in the Honduras. I showed how Halsey created the concept of 'violence' to help him form judgments about the way different machines affect the development of assemblages over time (Halsey, 2006). I also showed how Hillier introduced the concept of 'speculated potentials' to resolve the practical problems associated with identifying becomings during the strategic planning process (Hillier, 2007; 2011).

By borrowing these concepts and adapting them to my own line of enquiry, I designed a sustainable assessment method that treated the developing building as a 'complex space'. Design decisions were conceived according to judgements over 'speculating potentials' about 'what might become' of these spaces.

This broader approach to Deleuze's concepts helped me resolve problems similar to those identified and tackled by others. Such problems related to the way one perceives entities,[1] and the way one identifies and judges the way such entities might change over time.[2] As a result I was able to expand much further along the theoretical line between philosophy, theory and practice than the three studies reviewed in Chapter 3. Whilst the three studies do not state this intention so explicitly, this transition from philosophy to theory and practice is fundamental to all three studies. For Bonta it is a transition from philosophy into analytical practice, for Halsey it is a transition into environmental legal practice, and for Hillier it is a transition from philosophy to strategic planning practice. The developments made in this book show that Deleuze's philosophy can be translated into a theoretical framework for conceptualising the planning and development process. In other words, at this stage in the study, the task of making Deleuze useful was successful at a conceptual rather than a practical level.

Strategy 3: learning from other tools used in the field

Whilst these adapted Deleuzian concepts helped me resolve gaps in the theoretical framework, and demonstrated that Deleuze's concepts can be used to resolve issues in the conceptual development of planning, they were unable to answer some of the more practical questions related to the implementation of my outline proposals for a Speculative and Immanent Assessment Method. I identified gaps in the proposal related to the introduction and implementation of SIAM into practice. I resolved these issues by drawing on a method currently used in design assessment practice: the assessment method used as part of the UK's CDM Regulations. I selected this assessment because it reflected some of the Deleuzian arguments developed in my framework – namely, the idea that assessments should be undertaken during (immanent within) the design process, and the idea that assessments should be based on speculations (speculated potentials) rather than the properties of a completed design.

The sequence of development noted above reveals a steady shift from primary sources in the first strategy (Deleuze's seminal texts) to secondary interpretations of Deleuze's work in the second strategy (other Deleuze-inspired studies), and Deleuzian interpretations of practice-based sources in the third and final strategy (existing assessments used in practice). As such, the findings from Research Stage A demonstrate that, in order to make the transition from philosophy to a proposal for planning practice, one must be prepared to look outside of Deleuze's seminal texts and, indeed, outside of Deleuze's philosophy to other non-philosophical sources.

One reading of this observation would conclude that there are clear limitations to the practical usefulness of Deleuze's philosophy within the field of planning, because to use his concepts appears to demand that the method design shifts away from Deleuzian philosophy.[3] The problem with this reading is that it is centred on a historical account of Deleuze's philosophy. After all, one must remember that the primary sources against which this conclusion is derived are historical documents. Most of Deleuze's key texts were written in a period of thirty years between the 1960s and the 1990s, and many of the concepts within these texts were dependent on the latest advances in the fields he studied. The concept of becomings, for example, was derived from the discovery and study of singularities in several areas of science during the period that Deleuze was constructing his ontological stance (see Chapter 2).

Similarly, one must remember that, whilst Deleuze's concepts were created within different fields of knowledge, like science, such studies were far from exhaustive. His work does not include many spatial disciplines, such as planning. And whilst some of his texts show his willingness to move from theory to practice,[4] he does not explore this practical dimension in sufficient detail to offer a working alternative. This point re-iterates Campbell's and Richardson's argument discussed in Chapter 1 – namely, that abstract theory, especially theory with a strong link to philosophy, cannot be transferred directly into planning practice:

> Few, if any . . . philosophers had planning in mind as the context in which they intended their ideas to be applied. . . . Their horizons are therefore broader than a subsection of public policy making concerned with the creation of place and the mediation of space. This is not to dismiss this hugely significant body of work but to caution the appropriateness of direct translation.
>
> (Campbell, 2006: 93)

Thus, whilst the data shows that Deleuze's philosophy, as it appears in Deleuze's texts, is relevant only to the early theoretical stages, such conclusions are inevitable because these primary sources are historically situated documents written for very specific purposes. In other words, if one focuses on Deleuze's texts, then one cannot *expect* his concepts to be directly useful to the latest advances in disciplines that Deleuze did not study.

With this in mind, I support a second reading of the results from Research Stage A. This second reading suggests that, whilst the concepts created and used by Deleuze in his studies are not directly transferable into planning, they can be adapted to these fields or offer directions for creating concepts specific to these fields. Not only is this stance more reasonable; it also captures the spirit through which Deleuze's concepts were created. As noted in the introductory chapter, Deleuze notes:

> It is not the elements or the sets which define the multiplicity.[5] What defines it is the AND, as something which has its place between the elements or between the sets. AND, AND, AND – stammering.
>
> (Deleuze and Parnet, 2002: 26)

Thus, it seems that Deleuze anticipated the need for scholars to re-create his concepts and to add new concepts to respond to a given field (hence, the AND, AND, AND). Taking these points into account, I hold that the results from Stage A show that Deleuze's concept *can* be developed into a theoretical framework used to set out a new practical tool, but such concepts must be adapted to suit the unique demands of this tool.

Taking forward this conclusion, I turn now to the broader question as to whether the findings in Chapter 6 are specific to the case study of BRE assessments or whether they are sufficiently *generalisable* to apply to other essentialist tools and non-essentialist tools alike.

A theoretical framework for other essentialist assessments

In Chapter 2 I outlined a number of other planning tools constructed according to essentialist principles. These include multiple indicators, such as European Territorial Cohesion Indicators (ETCIs), the Index of Multiple Deprivation (IMD) and Housing Quality Indicators (HQIs); impact assessments, such as the Territorial Impact Assessments (TIAs) and the Heritage Impact Assessment (HIA); and design guides.

Looking at two of these tools in particular, one can see how essentialist principles underpin the multiple indicators used by the European Union to measure territorial cohesion across the member states (ETCIs) and impact assessment tools used to measure the territorial impacts of European and member state policy (TIAs). As in my review of the BRE assessments, a historical study of these two tools demonstrates the fundamental role played by the essentialist modelled concept. However, unlike BRE assessments, these two assessment tools are constructed using two different kinds of model: the 'tree model' and the 'storyline model' (Abrahams, 2014).

The former assumes that concepts like 'territorial cohesion' (also, the sustainable building) can be sub-divided into component dimensions. These dimensions are thought to capture all or some of the essential traits, characteristics or themes of the concept. Taken together, these dimensions provide one essential and universal meaning for the concept that can be used to measure or assess variations. As I showed in Chapter 2, BRE assessments like the Code for Sustainable Homes and the BREEAM New Construction assessment are good examples of this tree-model.

The 'storyline model', however, is not constructed on the assumption that a concept can be defined according to only one set of essential traits. Rather, this model is conceived as a collection of competing definitions deemed essential to many other policy concepts. The history of ETCI and TIA development is governed by the interplay between these two models. Whilst the tree model has dominated this development process to date, many of the latest assessment methods point to a hybrid approach, in which tree-model and storyline modelled definitions are included in the assessment's construction (Abrahams, 2014). This suggests that the theoretical framework developed in this study would apply only to areas of the ETCI and TIA constructed around tree-models and would be less applicable to areas of the ETCI and TIA methods developed around storyline models.

A further point for consideration relates to the third strategy of development used to adapt the proposal by drawing on assessment tools already used in practice (see above). In efforts to construct a non-essentialist form of assessment, I concluded that Deleuze's concepts were unable to respond to some of the most practical issues that arose during the study. These issues centred on a complicated relationship between design exploration and evaluation of the resulting proposals, as well as responses to the specific demands of a scheme, and regulation of the scheme against broader criteria. I responded to these issues by developing links to existing forms of assessment within the field. This showed that Deleuze's concepts alone could not meet all the demands expected of professional planning.

Thus, one might suggest that the success of a Deleuze-inspired planning tool is dependent on the kinds of regulatory tools already used in a given field of assessment, the context in which these tools operate and the purposes for which these existing kinds of tools were selected. If this study had been undertaken in an area of planning with less sophisticated forms of regulatory assessment, it is possible that the transition from a theoretical framework to a practice-focused tool would not have been made. In such instances, the conclusions to this study would have been that Deleuze's concepts are unable to make a valid contribution to planning practice.

This suggests that some of the Deleuzian concepts translated and adapted as part of this study would be applicable to other forms of assessment practice, particularly those constructed around essentialist 'tree models', whilst others would demand further adaptation. As such, I hold that only *some* of the findings from this study can be generalised to include other forms of essentialist assessment.

A theoretical framework for other kinds of assessment

Expanding this question further, one might ask how the theoretical proposals for SIAM relate to other assessments whose construction is less clearly, or not defined, by essentialist principles.[6]

In Chapter 2 I identified several different framings of Deleuze's ontology. Such framings include Deleuze's philosophy as a non-essentialist ontology (DeLanda, 2002), an object-orientated or machinic ontology (Bryant, 2011; 2014) and an event-orientated ontology (Massumi, 2013). Elsewhere I argue that these differences do not reveal contradictions but, rather, 'offer us a variety of starting points for navigating through Deleuze's key ideas and, by extension, through our own work' (Hillier and Abrahams, 2013: 5).

If one reflects on my review of other Deleuze-inspired studies as well as my own, it is highly likely that the limitations associated with identifying and judging becomings would be equally problematic to other studies, as would the practical problems surrounding the introduction and use of these concepts in planning and development practice. Thus, whilst an alternative framing might provide a different starting point, there is no reason to believe that framing Deleuze's philosophy in a different way would avoid some of the practical limitations in his concepts, such as the difficulties associated with the identification and assessment of 'becomings'. I hold, therefore, that the ideas developed in my theoretical proposals for SIAM offer general guidance for other studies focused on non-essentialist assessments in planning and development practice.

A practical and useful tool

As above, the following discussion will consider how the findings from the empirical stage of the research (Research Stage B) relate to the specific case of sustainable assessments and to assessments more generally.

A practical, useful sustainable assessment tool

Overall, respondents showed a great deal of support for the Deleuzian ideas used to underpin the proposals for a Speculative and Immanent Assessment Method – namely, that the assessment method should be undertaken during the design and development process; that it should be undertaken by those working throughout this process; and that it should be based on the contexts in which these decisions are made rather than a set of criteria determined in advance. However, this support was mostly identified with philosophical and theoretical strengths rather than its practicability, or its usefulness. The empirical evidence in Research Stage B suggests that, to make the theoretical tool into a practicable tool, further adaptations would need to consider its relationship with other tools and how actors in the planning and development process might engage with them in ways that have not been envisaged in the development of the SIAM proposal.

The evidence suggests that the principle motivations for adapting SIAM are to avoid manipulation by various actors in the planning/development process, to introduce accountability, and to allow for comparison between buildings and between a building and national government targets. The evidence suggests that meeting these demands would require the introduction of a fixed process or fixed set of criteria, and some form of quantitative measurement. In other words, to make these non-essentialist ideas of practicable use, they must be adapted to accommodate an essentialist model. Such changes would move SIAM away from its underlying principles – namely, to explore the possibility of a non-essentialist approach to assessing sustainability of a given building or urban design.

Two readings of the results from Research Stage B

There are two possible readings that can be drawn from this finding, each resulting in a different response to the second research question. In the following, I will review these readings in turn and assess their plausibility.

The first reading is based on the non-essentialist ontological frame outlined in Chapter 2. From this perspective, one would conclude that calls to re-integrate the essentialist modelled concept into SIAM show that Deleuze's concepts and the arguments on which they are based do not, and cannot, correspond to certain demands of planning and development practice. Such demands include the role played by planning to regulate different developments in order to reflect broader objectives.[7] As such, the evidence generated in the second research stage lends itself to the null hypothesis outlined in Chapter 4. In other words, Deleuze's concepts *cannot* be developed into a practicable tool used by actors in particular aspects of the planning and development process.

Whilst this reading of the evidence is a rational one, I believe that it places too much emphasis on the framing of Deleuze's philosophy rather than tackling the more pragmatic issue of usefulness underpinning this study. In Chapter 2, and as re-iterated above, I noted that Deleuze's philosophy has been framed in many different ways and that my use of the non-essentialist frame was a strategic one.

It is important to remember that Deleuze did not align himself with any single, ontological frame and so all efforts to collate Deleuze's concepts into such frames are interpretative. Selecting one frame over another, I argued, should, therefore, be based on its relevance to the problem under study and its ability to direct a field of enquiry.

A second reading of the evidence in Research Stage B would place more attention on the idea of usefulness rather than focusing on a single ontological frame. This reading would conclude that Deleuze's concepts *can* contribute to a practicable tool because respondents hold that Deleuze-inspired tools like SIAM do something useful that is not currently accommodated by essentialist tools like BREEAM – namely, that it better accommodates creative responses to specific problems arising in a given context. It would also suggest re-framing Deleuze's concepts to explain why respondents might favour essentialist *and* non-essentialist tools.

If we look at the data, it seems that respondents support these two kinds of tool because they serve different roles and relate to different actors in the planning and development process. The evidence suggests that essentialist tools are very effective when used to relate a specific building, its context and the actors that form a given project team to a much broader regional or national context, and to the actors that operate in that broader context (such as regional councils and UK government departments). The trade-off with such tools is that they are unable to take sufficient account of the unique considerations specific to a given context. Respondents suggested that non-essentialist assessments achieve the opposing benefits and limitations. They are very successful at accounting for the complexity of a specific scheme but unable to account for the work undertaken by actors operating at much larger scales and for their attempts to steer change towards broadly defined objectives.

If we think about this in Deleuzian terms, these conclusions suggest that these two different kinds of machines (essentialist and non-essentialist machines) filter different potentials to serve the operational demands of different assemblages. Thus, essentialist machines like BREEAM block potentials that are useful to the design-team assemblage but encourage potentials that are useful to the national government assemblage. Inversely, non-essentialist machines like SIAM block potentials useful to national government assemblages but encourage potentials useful to the design-team assemblage. This seems to reiterate Hillier's broader argument that the transcendental plane cannot be removed from planning altogether because it serves a functional role in planning, particularly in instances when planners need to enforce broader regulatory demands (see Chapter 2).

Halsey's study is equally relevant to this observation. If one assumes Halsey's stance on judgement – i.e. that a machine should be judged according to the potentials it encourages and blocks – then an entirely non-essentialist approach to assessment would result in a case of 'violence' (Halsey, 2006). This is because a system based exclusively on non-essentialist machines would prevent any potential for comparison across scales, to different groups of actors or the broader aims of these actors (such as reducing CO_2 levels nationally). The inverse is also true. A system based exclusively on essentialist machines also results in violence because it prevents any potential for responding to the specific demands of a given context.

A practical, useful assessment tool more broadly

Expanding the second research question further one might ask how the findings from the empirical stage of research (Stage B) relate to other assessments used in other areas of planning. Given the comments above, there is no reason to believe that the roles played by essentialist and non-essentialist tools would be any different when used in a different area of assessment. This is reflected in the data. Many respondents discuss essentialist assessment models by relating them to other essentialist models used in different areas of development and regulatory practice.

It is likely, therefore, that developing new tools for these other areas of planning practice would also demand that the researcher re-consider the use of DeLanda's essentialist/non-essentialist framework. This assertion would need to be explored in further research.

However, the question of usefulness does not necessarily preclude the continued presence and engagement with a variety of essentialist tools. This point is captured by R8, who notes that there is no way to remove (essentialist) models from the design process. Indeed, such models are tied into the way designers form their designs and the time pressures set out in the project programme. Rather, the role of SIAM, and, thus, one might imagine, all such Deleuze-inspired tools, is to make designers aware of these tools, what they do and do not do, and how they might play out differently. Such consciousness would be beneficial, R8 adds, because it would ensure that one would use models with a greater sense of awareness about their impacts:

> I think you will probably always get a stance that says this is how we do things – how we solve these problems. We only change it when it doesn't work. And I think you will still get that – which isn't necessarily a problem. But you will be more in control of determining the models.

This conclusion partly reflects some of the points drawn from Chapter 3 concerning the three Deleuze-inspired studies and directions for responding to the modelled concept. Like Bonta and Halsey, R8 suggests that modelled concepts are part of the 'complex spaces' considered in the design process. Like Bonta and Halsey, R8 also suggests that our efforts should be directed not at removing these modelled concepts but in developing an awareness about how these models affect the way assemblages come into being and adapt over time.

The role of 'grand' theory

These discussions concerning the usefulness of the Deleuze-inspired tool developed over the last few chapters lead me onto a much broader issue concerning the role played by 'grand' theory and the relationship between planning theory and planning practice. Above I concluded that in order to move from a philosophical framework to a planning theory and a practicable tool for planning practice, one must follow a sequence of strategies. Above I identified and discussed three strategies of development from Research Stage A and a further strategy from Research Stage B. These are summarised as follows:

Strategy 1: recreating Deleuze's concepts

Strategy 2: borrowing Deleuze-inspired concepts from other studies

Strategy 3: learning from other tools used in the field

Strategy 4: moving away from an ontological framing

As noted above, these four strategies reflect a progressive move away from established, theoretical understandings of Deleuzian philosophy to a case-specific, applicable understanding of Deleuze's philosophy. In its purest sense, therefore, this book has shown that grand theory like the one presented in Deleuze's philosophy has a role in theoretical analysis and debate but has a less obvious role in applied theory and practice. To expand this role, grand theory must be adapted to form case-specific theory.

These conclusions highlight the link between planning theory and planning practice and can be situated within broader debates within the field. The theory-practice relationship has been debated and theorised within the field of planning for decades (Lord, 2014). Whilst this debate has not been settled, there is broad consensus within the literature that these two areas of the discipline remain divided, and this division is problematic.

For the most part, the literature suggests that planning theory, rather than planning practice, is mostly responsible for the disconnection. For Breheny, the blame lies with theorists themselves whose work has been purposefully abstract, and misrepresentative of planning practice (Breheny, 1983: 114). Using Deleuzian terms, one explanation for this might be that this abstraction is the result a 'publication machine' that filters research becomings in favour of theoretical development rather than practice development.

For Binder, the division between theory and practice is not owing to misrepresentation or ignorance alone but in the direction taken by the academic community. Such theories, he argues, fail to

> provide a useful analysis of the planning system which can inform change . . . and are of real use to practitioners.
>
> (Binder, 2012: 221–222)

Thompson draws on a wealth of experience to offer his support to this stance. For Thompson, most academic publications make little attempt to consider how a planning theory might relate to practice, or whether practitioners will understand the concepts they allude to:

> The impression is of a small tribe of experts speaking to each other in strange tongues.
>
> (Thompson, 2010: 132)

This image of a 'small tribe of experts' speaking in 'strange tongues' could be applied to the Deleuzian community, whose work is often seen as impenetrable to many non-Deleuzian academics, and probably all the more impenetrable to practitioners who are unlikely to have engaged directly with philosophical constructs on a regular basis. With this in mind, this book simplifies and translates Deleuze's concepts into terms that may appeal to theorists and practitioners outside 'the small tribe of (Deleuzian) experts'. This has been achieved in several ways: by drawing links to my experiences in practice (Chapter 1, 2 and 7), by defining and tabularising the concepts used by other Deleuze-inspired theorists (Chapter 3) and by translating my own interpretation of Deleuze's concepts into terms more commonly used by actors working in planning and development roles (Chapter 6). Such efforts have been mostly successful, at least in methodological terms. A good indication of this success can be seen in the interviews discussed in Research Stage B. The two concepts 'speculation' and 'immanence' are not common terms in design and regulatory practice. However, none of the respondents interviewed struggled to understand these Deleuze-inspired concepts or questioned their use in a tool intended for practice.

Whilst language presents an issue in the theory-practice debate, Poulton suggests another reason for planning theory's estrangement from planning practice. Planning theory, she argues, is 'lost', with no clear idea as to its purpose or role in the discipline (Poulton, 1991: 225). Building on this claim, Poulton identifies four directions for subsequent research into planning theory. These include attempts to theorise the 'aims' of planning; attempts to theorise the procedures used by practitioners and how such procedures can be improved; attempts to develop 'theories of social criticism' whose aims are to critique and discuss the functions specific to planning; and 'positive theories' whose aims are to explain the activity of planning and how it evolved into its current state (Poulton, 1991: 230). Neuman captures these four roles for planning theory in much simpler terms, arguing that planning theory can be used as explanation, prediction, justification and normative guidance (Neuman, 2005; see also Yiftachel, 1989; Lord, 2014).

According to Neuman, whilst explanatory theory accounts for the majority of theoretical studies within the field, it is normative theory that holds 'the most promise'. For Neuman, it is not sufficient for theory to remain an explanatory endeavour either by 'describing practice' or by 'identifying pitfalls' (Neuman, 2005: 137). Rather, he argues, theory should consider what planning practice should do, how it should operate and how it should respond to issues that are of immediate concern to practitioners (Neuman, 2005: 137). Similar calls for substantive engagement with issues facing practice are equally developed in Harrison's review (2014). I believe that my attempts to translate philosophy into a practicable tool provide support to these arguments for normative practice.

As noted above, an important concern in this book has been about translating theory into practice, not only in terms of the language used but also about its relevance and its usefulness. The idea of a theory-to-practice translation has been strongly debated by a number of key figures in the field. For the planning theorist

Ernest Alexander, translation is one of two ways for planning theory to form the basis of practical action. Translation, he notes, has been particularly successful in the natural sciences, with few such examples within the social sciences. Building on this observation, Alexander asserts that, if planning theory is to affect practical action, then it is more likely to occur through a second route – i.e. through enlightenment. This indirect link between theory and practice, he adds, accounts for the perceived 'gap' noted in the discipline, which, he concludes, will remain 'unbridgeable' as long as the planning community continue their search for a translatable planning theory (Alexander, 1997: 4/5).

Alexander's call for planning theorists to focus on theoretical enlightenment of practice rather than translating theory into practice runs against the aims, and conclusions, of this study. Contrary to Alexander's assertion, the evidence from Research Stages A and B suggest that translation *is* possible, as long as one is prepared to accept that the theory one aims to translate will need to be adapted to meet the specific demands of the field of enquiry. By extension, this study suggests that it is by adapting abstract concepts in this way that one might establish a 'bridge' between planning theory and practice.

My conclusions regarding translation from planning theory to practice are supported by other planning theorists and planning practitioners. The idea that translating general theories into practice must also account for specific contextual issues is reflected in a number of academic papers ranging from the 1970s to the late 2010s. Scott and Roweis, for example, support attempts to identify abstract normative planning theories but argue that any such attempts should ask, 'under what specific social circumstances will any given imperative be relevant and applicable in practice?' (Scott and Roweis, 1977: 1099).

March makes a similar point when he notes that the contribution made by planning theory is dependent upon specific institutional circumstances. For March, it is important that the theory, and those using this theory, are aware of this relationship and take actions to ensure that applying the theory does not lead to unintended outcomes (March, 2010: 108). Similarly, Moroni suggests that certain kinds of theory are relevant to specific kinds of practice and this relationship should be explored and developed accordingly (Moroni, 2010: 138).

These comments offer broad support to the conclusions drawn from Research Stages A and B. More specific support can be seen in Friedmann's review of the translation process. In this review, Friedmann identifies three key lines of development within this process: considering how a given philosophy might relate to a specific area of practice; adapting normative practices to respond to a specific context; and translating knowledge and ideas from other fields into planning (Friedmann, 2008). Whilst Friedmann does not suggest that these developments should follow a particular order, one can see parallels with the four strategies of development drawn from my case study of SIAM and as set out above.

This brief review demonstrates that the conclusions I raise about translating Deleuze's philosophy into practice are echoed in debates about the role of theory-building in planning. It also suggests that these results might contribute to these on-going debates within the field.

The role of judgement in planning

Moving further along the line from theory to practice, I would like to briefly consider a second broad theme emerging from this study. This theme concerns the role by judgement in professional planning practice.

This book has highlighted two different ways of thinking about and making judgements in planning. The first form of judgement is captured in the arguments presented by Fischer, who suggests that planning judgements should be made in the name of a 'higher authority' both as the 'public good' and as the high-level governmental structures that enforce this 'public good' (Fischer, 2003; see Chapter 1). This view of judgement in the name of a higher authority has long since been an integral part of the planning profession (Campbell and Marshall, 2002; see also Meyerson and Baufald, 1955; Gower Davies, 1972). Klosterman notes:

> The profession's political roots in progressive reform were reflected in arguments for planning as an independent 'fourth power' of government promoting the general or public interest over the narrow, conflicting interests of individuals and groups.
>
> (Klosterman, 1985: 9–10)

Despite contestation within the field (Evans, 2007; Lowri, 1994; Moroni, 2004), many attempts have been made to protect this image of judgement (Klosterman, 1980; Alexander, 2002; Campbell and Marshall, 2002). Such efforts have not been limited to theoretical studies. From his ethnographic study of planning practice in Southern England, Malcolm Tait found that the ideological belief in a planning professional making judgements in the 'public interest' and to 'uphold public values' were key aspects of professional decision making (Tait, 2011: 165; see also Howe, 1994). Such conclusions are equally evident in some of the responses made by respondents in the empirical section of this book.

The second form of judgement in planning is captured in my proposals for a Speculative and Immanent Assessment Method. Rather than appealing to a higher authority and a transcendental value created by this authority (such as the sustainable building), SIAM looks to form immanent judgements based on an evaluation of the potentials to *become sustainable* particular to a given set of entities at a given moment in the design process. These proposals for a new approach to judgements in planning reflect Deleuze's calls for a shift away from assessments based on a transcendental image of the 'good' or 'bad' decision and towards an immanent form of evaluation that 'requires no appeal to transcendent criteria' (Smith, 2006). As Deleuze notes in *Cinema II*,

> It is not a matter of judging . . . in the name of a higher authority, which would be the good, the true; it is a matter, on the contrary, of evaluating every being, every action and passion, even every value, in relation to the life which they involve . . . [i]mmanent evaluation, instead of judgement as transcendent value.
>
> (Deleuze, 1989: 136)

This shift in judgement is supported by key figures in planning literature calling for assessments that place greater emphasis on the professional judgement of the planner as an expert in the field (Owens, 1994; Campbell, 1998; Vigar, 2012). These calls are also reflected in the empirical data discussed in Chapter 8. A number of respondents felt that judgements should be made in response to the context of the project and should rely on evaluations made *within* this context.

Thus, it seems that a method for making Deleuze-inspired judgements in planning can exist and would be supported by those working in academia and practice. But it also suggests that establishing this method as an alternative to the existing approach to judgement would demand a change in the way one thinks about planning and in the way planning relates to transcendental notions of the 'public good' and to higher administrative authorities.

So how might this shift come into being? Planning literature provides us with two possible directions for making this leap. The first of these is captured in Rydin's study of emerging, realist planning theory (2014). According to Rydin, realist ontological frameworks, like the one presented by Deleuze, point to three new challenges for planning practice:

> First, there is the de-centring of the planner as a key social actor from a central position within governance networks, enabling communication. . . . [Secondly] the materiality of planning practice, the mediation of the material elements within planning practice, the role of material artefacts, and the re-emergence of the physical site and environment as a focus. . . . Thirdly, the espousal of a flat ontology would seem to refute the significance of scale and of a cascade or nested hierarchy of strategy formulation at these different scales, from national through regional to local.
>
> (Rydin, 2014: 593–594)

Thus, for Rydin, establishing a realist approach to planning (such as those derived from Deleuzian realism) will be possible if this emerging paradigm continues to develop in planning theory and the field is able to provide practicable solutions to these new challenges. In other words, the shift towards Deleuzian judgement in planning will be possible only if his philosophy becomes further embedded in the field.

A second opportunity for making this transition to immanent assessment methods like SIAM can be seen in Purcell's proposals for bringing Deleuzian philosophy into planning. For Purcell, changing planning to better reflect Deleuze's ontological stance requires clear and directed revolutionary action (Purcell, 2013). For Purcell, it is only by constantly avoiding capture by the state machine that one can retain the spirit of Deleuze's creative approach. In this appraisal, SIAM would act as one of many lines of flight breaking through established processes and changing before it was adopted as a practice-wide tool.

These two proposals provide different images for, what Purcell terms, a 'new land' of planning: one in which Deleuze-inspired tools for making judgements in planning become established parts of planning practice, and one in which such tools are always on the edge of established planning practice. In terms of Deleuze's philosophy, these two directions come with their own risks. For the former, it is the risk that establishing such tools into mainstream practice will turn Deleuze's philosophy into a referential model. Whilst for the latter, the risk is that Deleuze's philosophy becomes an ideology that is unable to make an effective change outside of academic circles. I believe that finding ways to avoid these risks whilst making Deleuze as useful to planning practice as possible sets out the most significant challenge for the future of Deleuze-inspired theory and practice.

Notes

1 Either as a complete object with properties that can be used to define it (relations of interiority) or as an incomplete assemblage related to other assemblages (relations of exteriority). These differences are presented in Chapter 2.
2 That is, an assemblage's 'becomings' (potentials to become something else).
3 It is beyond the scope of this book to determine whether this observation is particular to Deleuze's philosophy or whether other philosophies can be adapted to practice without demanding similar shifts (see Chapter 10).
4 Such as his critique of Freudian psychoanalytical techniques.
5 The concept of the 'multiplicity' is explained and discussed in Chapter 3.
6 The BRE's Sustainability Checklist method provides a good example of this kind of assessment. The evidence from Research Stage B suggested that the construction of the Checklist method is less clearly or not defined by essentialist principles.
7 Such as regulation for the 'public good'.

10 Conclusions

I would like to conclude this book by reflecting on the contributions made by this study into Deleuze and planning, but also consider its limitations and how they might be overcome in further research.

Contributions of the study

Principle contributions

The principle contribution made by this book concerns the translation of Deleuze's seemingly obscure concepts into a new, practicable tool that can be used by actors in the design, development and regulatory process. As noted in the introduction, making this transition is difficult because it requires planning theorists to translate Deleuze's work across three different bodies of knowledge: philosophy, planning theory and planning practice. Whilst a number of theorists support this transition on the basis that such links might create 'new practices and tools to increase the effectiveness of [the discipline]' (de Roo et al., 2012: 20), few have successfully explored this transition in detail.

Understanding and expanding Deleuze's concepts

This gap in knowledge is partly owing to the difficulties associated with Deleuzian terminology. Without a strong foundation in Deleuzian scholarship, Deleuze's concepts can appear opaque and abstract. Over the years, several attempts have been made to resolve this issue. Adrian Parr's edited collection of dictionary definitions is one of the most notable examples (Parr, 2005). However, whilst such publications provide a useful reference for philosophers, it does not fully reflect the demands of an audience working in other fields, such as planning. This is evident in the dictionary definition entry for the concept 'becoming', for example, which reads:

> Deleuze uses the term 'becoming' (devinir) to describe the continual production (or 'return') of difference immanent within the construction of events,

> whether physical or otherwise. . . . Deleuze's theory holds that things and states are *products* of becomings.
>
> (Parr, 2005: 22)

As this quotation suggests, the concept of 'becoming' is defined as part of a network of other Deleuzian concepts such as 'difference', 'immanence', 'events' and 'assemblages'.[1] Not only do these definitions demand prior knowledge of Deleuzian terminology; they also provide little indication as to their use outside philosophy, or explain why their definitions do not correspond to the definitions used by others. This book helps resolve this gap by defining such concepts according to a real-world example in the design and regulatory process (Chapter 2), defining and comparing the way others have used these concepts in other spatial disciplines (Chapter 3) and showing how these definitions relate to and inform a specific area of planning practice (Chapters 6 and 8).

In doing so, this study contributes to a community of spatial theorists interested in Deleuze's ideas but who have experienced difficulties trying to make sense of concepts like 'assemblages', 'becomings' and 'multiplicities'. By presenting real-world examples of these concepts and showing how, why and when Deleuze's concepts might be adapted to inform very practical problems, it is hoped that planning theorists will be more open to using Deleuze's philosophy and derive a more realistic appreciation of what such theories can achieve.

Understanding and expanding other Deleuzian studies

My efforts at making Deleuze's philosophy more accessible to a non-philosophical audience have not focused on Deleuze's seminal texts alone. As part of this study, I have reviewed three of the most developed Deleuze-inspired studies undertaken in the spatial disciplines. These studies are Mark Bonta's study of land use in Honduras; Mark Halsey's study and proposals for environmental damage within the Goolengook forest, Australia; and Jean Hillier's proposals for a Deleuze-inspired approach to strategic planning. Whilst these studies have been championed by other scholars in the field (see Chapter 3), this literature fails to critically engage with these studies and, particularly, how they translate Deleuze's philosophy in ways that made it relevant to specific issues within their respective disciplines. This gap in knowledge leaves many key questions unanswered. Such questions include: why was one Deleuzian concept selected over another? How were these concepts adapted to respond to the specific demands of the field? How did these concepts feed into a broader theoretical framework? What were the gaps in this framework, and why did these gaps occur?

By tackling these questions through a detailed, systematic review of the three studies, I have tried to make a valuable contribution to all three bodies of knowledge and Deleuzian scholarship more broadly. My efforts to capture the theoretical frameworks in tables provide subsequent geographers, environmentalists and strategic planners with a clear starting point to understand and expand on the contribution made by each. As part of this analysis, I critically reviewed the contributions and gaps within the respective frameworks and explored opportunities for resolving these gaps to inform studies undertaken elsewhere. In Bonta's study, I suggest that a more refined understanding of the concept 'complex space' could contribute to other spatial studies such as 'concept spaces' in ANT (Callon and Law, 2004) and Healey's proposals for 'relational space' (2006). Halsey's concept of 'judgement', I argued, could contribute to the judgements made in planning policy (this is also discussed in Abrahams, 2014). In my review of Hillier's study, I showed that Hillier's multiplanar theory distinguishes three influences leading to the creation of a planned or designed assemblage: speculated potentials directed towards empty signifiers (or singularities); the un-speculated potentials that emerge over time; blueprints for a specific geographical location or for a specific area of design (such as established layouts for WCs, houses or cul-de-sacs). Understanding these processes, I argued, could provide the basis for a new Deleuzian method of plan-making used by planners, masterplanners and architects alike.

Not only did I identify opportunities within each study for further development; I also brought these conclusions together to identify generalisable conclusions about the measures needed to make Deleuze's philosophy useful to the spatial disciplines. I showed that all three studies demonstrated that making Deleuze's concepts useful meant adapting and expanding them to respond to the specific demands of problems identified in a field of knowledge. I showed that failures to develop a viable proposal for engagement corresponded to areas of the theorist's theoretical framework where such adaptations and expansions had not been made.[2]

Impacts on the theory-practice 'gap'

These arguments set out a foundation for the case study in Part 3. This study focused on an area of assessment dominated by a 'universal method' formulated by the BRE[3] (see Chapter 4). The resulting case study was constructed into two stages. In Research Stage A, I set out a theoretical framework and proposals for a Deleuze-inspired assessment tool, which I named the 'Speculative and Immanent Assessment Method' (SIAM). In Research Stage B, I tested this proposal through empirical enquiry.

Taken together, these two stages demonstrated that Deleuze's abstract concepts *can* be translated into a practicable tool deemed useful to actors working in the development and regulatory process. Drawing on the results of the case study, I have identified four strategies needed to make this transition from Deleuzian philosophy to planning theory and planning practice:

Strategy 1: recreating Deleuze's concepts

Strategy 2: borrowing Deleuze-inspired concepts from other studies

Strategy 3: learning from other tools used in the field

Strategy 4: moving away from an ontological framing

These strategies show that the transition from philosophy to practice demands a steady shift away from primary source, to secondary Deleuzian sources, to sources with no explicit link to Deleuze's philosophy. In the last chapter I argued that this shift away from Deleuze's philosophy did not necessary imply that Deleuze's concepts were less useful as the case study developed. Rather, I argued that this finding revealed the importance of individual interpretation, and the theorist's ability to re-create concepts in new ways to respond to field specific demands. This reflects Richardson's and Campbell's point that abstract theory cannot be drawn directly into practically motivated disciplines like planning. I believe that these conclusions offer subsequent theorists a realistic expectation about *what* can be achieved by this Deleuze/planning link, as well as *where* and *when* this link might prove useful and effective.

A second set of contributions

The 'modelled concept' in planning

One of the further contributions made by this book concerns the kind of tools used in planning practice. In Chapters 1 and 2, I identified a common format used to construct impact assessment methods, indicators and design guides. This format, I demonstrated, was based around a concept like the 'sustainable home' or the 'sustainable building', defined according to its essential traits in order to provide a referential model for assessing a variety of different designs, projects or situations. Whilst there has been a great deal of debate surrounding the impacts of essentialism, and efforts to challenge its role in planning practice (see Harper and Stein, 2006), the literature does not offer a common vocabulary for discussing these essentialist models. This study contributes to this body of knowledge by offering the term 'modelled concept' into these debates and by showing the role played by 'modelled concepts' in normative planning practice.

Two kinds of tool and their uses

Like pragmatists such as Stein and Harper, I argued that 'modelled concepts' do not offer a fair and reasonable reflection of the difficult decisions made during the design/development process, and that they are unable to take sufficient account of differing contextual factors (Harper and Stein, 2006; Stein and Harper, 2012). On this basis, this study develops a non-essentialist tool that I term the 'Speculative and Immanent Assessment Method' (SIAM). Unlike proposals made by Stein and Harper, this tool is not limited to a human-centric (non-realist) approach. Rather, it is designed to take account of human and non-human assemblages that form the project as a whole, and to focus on the pre-actual becomings that make such assemblages possible.

Whilst the 'modelled concept' method noted above provides a new framework for discussing methods constructed using essentialist principles (such as BREEAM), the proposals for SIAM offer a new framework for discussing methods based on non-essentialist principles. The CDM assessment method discussed in Chapter 6 and the BRE sustainability Checklist identified through empirical enquiry in Chapter 8 are both designed to prioritise speculations over properties, and are undertaken during the design process (immanent) by different members of the project team rather than by an external assessor.

This distinction between essentialist assessments designed around 'modelled concepts' and non-essentialist tools designed as a Speculative and Immanent Assessment Method offer a strong contribution to assessment methodology. But the study also reveals problems in the simple dualism between a problematic essentialist method and a desirable non-essentialist method seen in assessment literature (see Chapter 2). In this study I have shown that modelled concept-based assessments and speculative and immanent assessment methods serve different roles and support different potentials (becomings) favoured by different assemblages operating at different scales. As noted in the last chapter, the evidence from this study suggests that essentialist tools are very effective when used to relate a specific building, its context and the actors that form a given project team to a much broader, regional or national context, and to the actors that operate in that broader context (such as regional councils and UK government departments). The trade-off with such tools is that they are unable to take sufficient account of the unique considerations specific to a given context. Non-essentialist assessments achieve the opposing benefits and limitations. They are very successful at accounting for the complexity of a specific scheme but unable to account for the work undertaken by actors operating at much larger scales and for their attempts to steer change towards broadly defined objectives. As such, this book makes a valuable contribution to the way one thinks about and discusses assessment methodology, as well as the way one considers their application in different instances.

Impacts on the role of judgement in planning

The more nuanced understanding of essentialist and non-essentialist methods described above contributes to the way one thinks about the role played by judgement in planning. As noted in Chapter 9, judgements in planning are often characterised as an appeal to a 'higher authority' rather than to an 'immanent evaluation' in which judgements are made by considering 'every being, every action and passion, even every value, in relation to the life which they involve' (Deleuze, 1989: 136). This study suggests that there is a practical as well as academic demand for developing immanent forms of judgement through methods like SIAM. It also suggests that such tools might pave the way to a 'new land': one that is based on Deleuzian conceptions of 'the real' and the creations formed *within* the real.

Other contributions

The contributions outlined above reflect the contributions outlined in the introductory chapter. However, during the course of this book, I have also identified a number of other contributions not anticipated at the outset.

Transcendental empirical methods

These contributions centre on the difficulties associated with developing a research design suitable for testing the SIAM proposals through empirical enquiry. Such difficulties stem from an area of Deleuze's epistemological stance termed 'transcendental empiricism'. As an extension of Deleuze's ontological frame, this stance argues that we, as researchers, are actualised assemblages operating in and around an actual realm. Given that our capacity to perceive the real is limited to this realm, Deleuze suggests that the researcher is unable to identify areas of reality that extend into the pre-actual – namely, the virtual realm and the processes of actualisation (see Chapters 2 and 7). My review of this epistemological assertion revealed a difficult question that has received little attention in research methods literature: *how does one design empirical research to account for a pre-actual area of reality that cannot be accessed by our sensory observations?*

To respond to this question, I undertook a detailed review of efforts to develop a Deleuzian research method for the social sciences (Coleman and Ringrose, 2013). One of the most problematic issues surrounding this work concerns a common distinction between data generation and data analysis. All the studies in Ringrose and Coleman's edited volume drew on Deleuze's concepts to set out proposals for new data analytical methods but not for data generation methods. Instead, each study used established, sensory-based methods for generating qualitative data. This, I argued, did not reflect Deleuze's arguments for transcendental empiricism and, as such, were not suitable to a Deleuze-compatible approach. My critical review of these Deleuzian research methods introduces new lines of debate within this new and expanding body of literature.

Critical realism and Deleuzian realism

In the absence of an established set of Deleuzian empirical methods I looked to comparable efforts made by critical realists. This review compared the onto-epistemological stance taken by these two forms of realism and identified a number of similarities and differences. This brief review offers some support to theorists working in one, or both, of these areas and reveals new opportunities for theorists to move between the two.

A speculative method of data generation

To respond to the problems associated with existing Deleuzian research methods, and the differences between critical realism and Deleuzian realism, this study draws on some of the ideas presented in Hillier's proposals for strategic planning as reviewed in Chapter 3 – namely, the concept of 'speculated potentials' (becomings) and the use of 'empty signifiers'. This study uses these concepts as part of a research method. As such, it provides new opportunities for developing Hillier's multiplanar theory in connection with the efforts made by Deleuzian research methodologists like Ringrose and Coleman.

Limitations of the study

The contributions made by this book are not without their limitations. Some of the most significant are identified and reviewed below.

The limits of an essentialist/non-essentialist frame

One of the most interesting findings of this study concerns the fourth strategy in the transition between Deleuze's philosophy and planning practice: the shift away from the essentialist/non-essentialist ontological framing. In the last chapter, I discussed why shifting away from the essentialist/non-essentialist, ontological frame did not invalidate the claim that Deleuze's concepts *can* be made useful to practice.

However, the fact that this shift was necessary demands further reflection about the framing and elaboration of the problem directing this book – namely, the idea that essentialist assessment methods are problematic because they do not take sufficient account of the complex decisions made during the design process.

The evidence suggests that the use of essentialist principles to model concepts like 'the sustainable home' or the 'sustainable building' *do* block opportunities to respond to complex issues specific to a site or/and project. However, the evidence shows that these same essentialist assessment tools (machines) also create opportunities for linking a specific design to broader national targets. Inversely, non-essentialist tools (machines) encourage opportunities to respond to the specific demands of a scheme but prevent links between different schemes, and to broader national targets. Another point raised in the empirical study is that essentialist models are used in all aspects of design, development, assessment and regulation. They are not, as suggested in Chapter 1, limited to assessment practice.

Thus, it seems that the problem for which Deleuze can be made useful is not about essentialism *per se*. Rather, it is about the way one *uses* essentialist and non-essentialist tools (machines), as well as one's awareness about what these tools do, and how they affect what other actors or tools do. These conclusions provide a much deeper insight into my own experiences within a specific area of practice, as well as insight into the experiences used to support the non-essentialist studies by pragmatists like Harper and Stein (see Chapter 2). This study offers some support to Harper and Stein's arguments for a non-essentialist approach to planning, but it also supports the broader point made by discourse analysts like Feinstein, who suggests that one must take account of the broader forces acting on practice. In other words, it suggests that it is only by moving in and out of a specific field or situation that one can better understand the role played by essentialism in planning, and, thus, to identify instances in which essentialism is problematic, and where it is beneficial. This conclusion provides a further interpretation of the Deleuzian message quoted at the beginning of this book:[4]

> partir au milieu, par le milieu, entrer et sortir, non pas commencer ni finir. . . . Ils ont su faire une pragmatique. C'est que le milieu n'est pas du tout une moyenne, c'est au contraire l'endroit où les choses prennent de la vitesse.[5]
>
> (Deleuze and Guattari, 1980: 37)

Limits of the empirical 'test'

Another limitation of the research concerns the method used to test the practicability of the SIAM proposal. The observations, comments and speculations made by the respondents in Research Stage B were hypothetical. As such, any conclusions about the practical viability of SIAM are limited to the respondents' imagination and their experiences of other assessment methods. With this in mind, I believe that piloting SIAM in a 'live' project could offer an additional, more robust test of its practical viability. This 'live' project could be studied using ethnographic methods[6] intended to reveal the way SIAM is interpreted and adapted into the design process. To do this, future research would need to navigate a number of difficulties associated with, but not limited to, case selection, access to decision making, the risk of bias drawn from the context of the ethnographic study, and limitations on the scope of actors included.

Opportunities for future research

Based on the limitations discussed above, I have identified four directions for future research into the relationship between Deleuze and planning as well as the development of a Speculative and Immanent Assessment Method:

1. To return to the milieu outlined in Chapter 1 and map as many essentialist and non-essentialist machines in the design, assessment and regulatory process as possible. Such studies would look to better understand how these machines operate/filter potentials and how these machines relate to different actors.
2. To undertake a parallel study into other forms of assessment in a different field of planning, such as the development of Territorial Impact Assessments or European Territorial Cohesion Indicators. This study would ascertain which of the adapted, re-created and expanded concepts in my theoretical framework can be used to inform an assessment tool specific to this field.
3. To undertake a second set of interviews with professionals working for private housing developers, the BRE and different levels of policymaking/implementation. The results from these interviews could be combined with the results noted in this study to offer a more detailed response to the issue of practicability.
4. To test SIAM using a 'live' project. This pilot scheme for SIAM would offer a unique insight into the practical limitations and advantages offered by this method over existing BRE methods. Given the comments raised in the empirical stage of the research, I believe this pilot scheme would be best undertaken on a mixed-use scheme formed from private commercial and not-for-profit schemes.

Final remarks

As a concluding note I would like to return to the idea of the schiz used to introduce this book: the idea that this study looks to satisfy Gareth the academic and

Gareth the practising architect, Gareth the abstract theorist and Gareth the pragmatist. Like all studies, it remains partial. It suggests that Deleuze's philosophy is an abstract machine whose potentials must be drawn towards the actual as well as stretched from the actual back to the virtual. As Deleuze notes, it is we 'the conceptual personae' that must make this happen. The question we must ask ourselves is not whether Deleuze can be made useful but whether we are prepared to act in this role, to seek out new connections, new opportunities to analyse *and* engage in the processes of actualisation.

Notes

1 Assemblages being the 'things' produced by becomings (see Chapter 2).
2 Halsey's failure to adapt and expand Deleuze's concept of 'becoming' to meet his proposals for judging a machine provides a good example of this observation.
3 Both as the BREEAM suite of assessments and as the Code for Sustainable Homes assessment.
4 This quote was originally interpreted as a method for undertaking this study rather than as a method for understanding and creating planning tools more broadly.
5 I translate this as follows: Start from a specific field, by that field, enter and leave it without starting or finishing. . . . This is how one acts pragmatically. This field (also translates as middle) should not be understood as an average. It is quite the contrary, for it is in this place that things achieve the greatest speed.
6 Adapted to reflect the specific demands of the study and to take account of Deleuze's epistemological stance.

Bibliography

Abrahams, G. (2014). What is territorial cohesion? What does it do?: essentialist versus pragmatic approaches to using concepts. European Planning Studies, 22/10: 2134-2155. http://www.tandfonline.com/doi/full/10.1080/09654313.2013.819838#.UlQ9Xn9wbmI

Abrahams, G (2013) 'Engaging in complex spaces: Drawing on two Deleuze-inspired voices' Paper presented at 6th Deleuze Studies International Conference in Lisbon, Portugal, 2013. Available online: http://deleuze2013.fc.ul.pt/ [Last accessed: 07/08/2013]

Abrahams, G; Hillier, H (2016) 'The essentialist ghost in the planning machine: Unlocking potentials in Western Australia's R Codes' (paper currently under review)

Alexander, E R (1997) 'A mile or a millimeter? Measuring the "planning theory–practice gap"' *Environment and Planning B: Planning and Design* 24: 3–6

Alexander, E R (1998) 'Where do we go from here?: Evaluation in spatial planning in the post-modern future'. In Lichfield, N; Barbanente, A; Borri, D; Khakee, A; Prat, A (Eds.), *Evaluation in Planning: Facing the Challenge of Complexity*. Dordrecht, Netherlands: Kluwer: 355–374

Alexander, E R (2002) 'The public interest in planning: From legitimation to substantive plan evaluation' *Planning Theory* 1(3): 226–249

Alexander, E R (2006) 'Evaluations and rationalities: Reasoning with values in planning'. In Alexander, E R (Ed.), *Evaluation in Planning: Evolution and Prospects*. Aldershot, UK: Ashgate: 39–52

Alexander, E R; Faludi, A (1989) 'Planning and plan implementation: Notes on evaluation criteria' *Environment and Planning B: Planning and Design* 16: 127–140

Allen, J (2011) 'Powerful assemblages?' *Area* 43(2): 154–157

Ansaloni, F; Tedeschi, M (2015) 'Ethics and spatial justice: Unfolding non-linear possibilities for planning action' *Planning Theory*. E-publication ahead of print 25 June. DOI: 10.1177/1473095215591676

Atkinson, A R; Westall, R (2010) 'The relationship between integrated design and construction and safety on construction projects' *Construction Management and Economics* 28(9): 1007–1017

Aubree, A (date unknown) 'Presentation on BREEAM International'. Available online: http://www.fsr.is/lisalib/getfile.aspx?itemid=4690 [Last accessed: 18/03/2014]

Baldwin, R; Yates, S; Rao, S; Howard, N (1998) *BREEAM 98 for Offices: An Environmental Assessment Method for Office Buildings (BR 350)*. Berkshire, UK: BRE Publications

Barnett, H (1999) 'The land ethic and environmental crime' *Criminal Justice Policy Review* 10(2): 161–192

Barrett, H C (2001) 'On the functional origins of essentialism' *Mind and Society* 3(2): 1–30

Batty, M (1996) 'Modelling the city: Performance policy and planning – Bertuglia, CS, Clarke, GP, Wilson, AG' *Progress in Human Geography* 20(2): 260–262

Batty, M (2011) 'A generic framework for computation spatial modelling'. In Heppenstall, A J; Crooks, A T; See, L M; Batty M (Eds.), *Agent-based Models of Geographical Systems*. Berlin and New York: Springer Verlag: 19–50

Batty, M (2011a) 'Modeling and simulation in geographic information science: Integrated models and grand challenges' *Procedia – Social and Behavioral Sciences* 21: 10–17

Batty, M (2011b) 'Dynamics of land-use change and growth'. In Robinson, D (Ed.), *Computer Modelling for Sustainable Urban Design*. London, UK: Earthscan/James & James: 237–258

Batty, S E (2006) 'Planning for sustainable development in Britain: A pragmatic approach' *Town Planning Review* 77(1): 29–40

Baum, H S (1996) 'Why the rational paradigm persists: Tales from the field' *Journal of Planning Education and Research* 15(2): 127–134

Bell, J (2006) 'Charting the road of inquiry: Deleuze's humean pragmatics and the challenge of Badiou. *The Southern Journal of Philosophy* 44(3): 399–425

Benton, T (1998) 'Rights and justice on a shared planet: More rights or new relations' *Theoretical Criminology* 2(2): 149–176

Bevir, M; Rhodes, R A W (2003) *Interpreting British Governance*. London: Routledge

Bevir, M; Rhodes, R A W (2006) *Governance Stories*. Oxon: Routledge

Bhaskar, R (1998) 'General introduction'. In Archer, M S; Bhaskar, R; Collier, A; Lawson, T; Norrie, A (Eds.), *Critical Realism: Essential Readings*. London: Routledge: ix–xxiv

Biernacki, P; Waldorf, D (1981) 'Snowball sampling: Problems and techniques of chain referral sampling' *Sociological Methods and Research* 10(2): 141–163

Binder, G (2012) 'Theory(izing)/practice: The model of recursive cultural adaptation' *Planning Theory* 11(3): 221–241

Blaise, M (2013) 'Activating micropolitical practices in the early years: (Re)assembling bodies and participant observations'. In Coleman, R; Ringrose, J (Eds.), *Deleuze and Research Methodologies*. Edinburgh: Edinburgh University Press: 164–183

Blumer, H (1969) *Symbolic Interactionism: Perspective and Method*. Berkeley: University of California Press

Bolger, F; Stranieri, A; Wright, G; Yearwood, J (2011) 'Does the Delphi process lead to increased accuracy in group-based judgmental forecasts or does it simply induce consensus amongst judgmental forecasters?' *Technological Forecasting and Social Change* 78: 1671–1680

Bonta, M A (2001) *Mapping enredos of complex spaces: A regional geography of Olancho, Honduras*. Unpublished doctoral dissertation. Baton Rouge: Louisiana State University

Bonta, M A (2008) 'Book review: Deleuze and environmental damage: Violence of the text' *Geographical Review* 98(4): 573–576

Bonta, M A (2009) 'Taking Deleuze into the field: Machinic ethnography for the social sciences' Review essay in *Deleuze Studies* 3: 135–142

Bonta, M A; Protevi, J (2004) *Deleuze and Geophilosophy: A Guide and Glossary*. Edinburgh: Edinburgh University Press

Booher, D; Innes, J (2002) 'Network power in collaborative planning' *Journal of Planning Education and Research* 21(3): 221–236

Borri, D (1998) 'Linking planning to theory: Introduction'. In Lichfield, N (Ed.), *Evaluation in Planning: Facing the Challenge of Complexity*. Dordrecht, Netherlands: Kluwer Academic Publishers: 275–282

BRE (2011) 'BREEAM new construction: Non domestic buildings, technical manual. SD5073–2.0:2011'. BRE Global Limited

BRE (2012) 'BREEAM communities: Integrating sustainable design into masterplanning'. BRE Global Limited

BRE Global (2011) 'News from BRE: New developments to become more sustainable' Available online: http://www.bre.co.uk/news/New-developments-to-become-more-sustainable-750.html [Last accessed: 17/01/2013]

Breheny, M J (1983) 'A practical view of planning theory' *Environment and Planning B: Planning and Design* 10: 101–115

Brisman, A (2008) 'Crime-environment relationships and environmental justice' *Seattle Journal for Social Justice* 6(2): 727–817

Bristol City Council (2011) 'Bristol development framework core strategy: Adopted June 2011'. Available online: http://www.bristol.gov.uk/sites/default/files/documents/planning_and_building_regulations/planning_policy/local_development_framework/Core%20Strategy%20WEB%20PDF%20%28low%20res%20with%20links%29.pdf [Last accessed: 05/06/2014]

Bryant, L (2011) *The Democracy of Objects*. London, UK: Open Humanities Press

Bryant, L; Srnicek, N; Harman, G (2011) *The Speculative Turn: Continental Materialism and Realism*. Melbourne: re.press

Bryant, L B (2014) *Onto-cartography: An Analogy of Machines and Media*. Edinburgh: Edinburgh University Press

Bryant, L R (2008a) *Difference and Givenness: Deleuze's Transcendental Empiricism and the Ontology of Immanence*. Evanston, IL: Northwestern University Press

Bryant, L R (2008b) 'The transcendent and the transcendental'. Blog Entry on 07/12/2008. Available online: http://larvalsubjects.wordpress.com [Last accessed: 25/02/2014]

Buchanan, I (no date) 'Introduction to Deleuze Connection series'. Available online: http://www.euppublishing.com/series/delco [Last accessed: 08/05/2014]

Buchanan, I; Lambert, G (2004) 'Introduction: Deleuze and space'. In Buchanan, I; Lambert, G (Eds.), *Deleuze and Space*. Edinburgh: Edinburgh University Press: 1–15

Bygstad, B; Munkvold, B E (2011) 'Exploring the role of informants in interpretive case study research in IS' *Journal of Information Technology* 26(1): 32–45

Callon, M. (1986) 'Some elements of a sociology of translation: Domestication of the scallops and the fishermen of St Brieuc Bay'. In Law, J (Ed.), *Power, Action and Belief: A New Sociology of Knowledge*. London: Routledge & Kegan Paul: 196–233

Callon, M; Law, J (2004) 'Introduction: Absence – presence, circulation, and encountering in complex space' *Environment and Planning D: Society and Space* 22(1): 3–11

Campbell, H (1998) 'Acting on principle: Dilemmas in planning practice' *Planning Practice and Research* 13(2): 117–128

Campbell, H (2006) 'Just planning: The art of situated ethical judgment' *Journal of Planning Education and Research* 26: 92–106

Campbell, H (2012) 'Lots of words... but do any of them matter? The challenge of engaged scholarship' *Planning Theory and Practice* 13(3): 349–353

Campbell, H; Marshall, R (2002) 'Utilitarianism's bad breath? A re-evaluation of the public interest justification for planning' *Planning Theory* 1(2): 163–187

Cartwright, R L (1968) 'Some remarks on essentialism' *The Journal of Philosophy* 65(20): 615–626

CEC (1992) 'Directive 92/57/EEC of the European Union and of the Council of 24 June 1992 on the implementation of minimum safety and health requirements at temporary or mobile construction sites' *Official Journal of the European Communities* L 245, 26/08/1992: 6–22

Charmaz, K (2006) *Constructing Grounded Theory: A Practical Guide through Qualitative Analysis*. London: Sage Publications

Cole, D R (2013) 'Lost in data space: Using nomadic analysis to perform social science'. In Coleman, R; Ringrose, J (Eds.), *Deleuze and Research Methodologies*. Edinburgh: Edinburgh University Press: 219–237

Cole, R J (2005) 'Building environmental assessment methods: Redefining intentions' *Building Research Information* 33(5): 455–467

Coleman, R; Ringrose, J (Eds.) (2013) *Deleuze and Research Methodologies*. Edinburgh: Edinburgh University Press**Cowell, R; Martin, S (2002)** 'The joy of joining up: Modes of integrating the local government modernisation agenda' *Environment and Planning C: Government and Policy* 21: 159–179

Cowell, R; Owens, S (2002) *Land and Limits: Interpreting Sustainability in the Planning Process*. Oxon: Routledge

Cupples, J (2011) 'Wild globalization: The biopolitics of climate change and global capitalism on Nicaragua's Mosquito Coast' *Antipode* 44(1): 10–30

Czernich, N; Falck, O; Kretschmer, T; Woessmann, L (2011) 'Broadband infrastructure and economic growth' *The Economic Journal* 121: 505–532

Danermark, B; Ekstrom, M; Jakobsen, L; Karlsson, J C (2002) *Explaining Society: Critical Realism in the Social Sciences*. London: Routledge

Da Silva, A W L; Selig, P M; De Avila Leripio, A (2014) 'Strategic environmental assessment: One concept, multiple definitions' *International Journal of Innovation and Sustainable Development* 8: 53–76

Davoudi, S (2003) 'Polycentricity in European spatial planning: From an analytical tool to a normative agenda?' *European Planning Studies* 11(8): 979–999

DCLG (2005) 'Planning policy statement: Planning and climate change. Supplement to Planning Policy Statement 1'. West Yorkshire, UK: Communities and Local Government Publications. Available online: http://webarchive.nationalarchives.gov.uk/20120919132719/http://www.communities.gov.uk/documents/planningandbuilding/pdf/153119.pdf [Last accessed: 08/05/2014]

DCLG (2006) 'Code for sustainable homes: A step-change in sustainable home building practice'. Communities and Local Government Publications

DCLG (2010) 'Code for sustainable homes: Technical guide, November 2010'. London: DCLG

DCLG (2012a) '2012 consultation on changes to the Building Regulations in England: Summary of responses'. London: DCLG Publications

DCLG (2012b) 'National Planning Policy framework'. London: DCLG Publications

DCLG (2013) 'Housing standards review: Towards more sustainable homes'. London: DCLG Publications

DCLG (2014) 'Government response to the Environmental Audit Committee Report: Code for Sustainable Homes and the Housing Standards Review: Presented to Parliament by the Secretary of State for Communities and Local Government by Command of Her Majesty'. March 2014. Printed in the UK by the Williams Lea Group

Dean, G C (1988) 'The domestic environment of the rural subsistence family in Northern Honduras' *International Journal of Sociology of the Family* 18(2): 165–186

Degeling, P (1995) 'The significance of "sectors" in calls for urban public health intersectionalism: An Australian perspective' *Policy and Politics* 23(4): 289–301

DeLanda, M (1997) *A Thousand Years of Nonlinear History*. Cambridge, MA: MIT Press, Swerve Editions

DeLanda, M (2002) *Intensive Science and Virtual Philosophy*. London: Continuum Press

DeLanda, M (2006) *A New Philosophy of Society: Assemblage Theory and Social Complexity*. London: Continuum Press

DeLanda, M (2010) 'Cities and nations'. In Hillier, J; Healey, P (Eds.), *The Ashgate Research Companion to Planning Theory: Conceptual Challenges for Spatial Planning*. Surrey, UK: Ashgate: 251–272

DeLanda, M; Protevi, J; Thanem, T (2003) 'Deleuzian interrogations: A conversation with Manuel DeLanda, John Protevi and Torkild Thanem'. Available online: http://www.dif-ferance.org/DeLanda-Protevi.pdf [Last accessed: 12/01/2013]

Deleuze, G (1986 [1983]) *Cinema 1: The Movement-image*. Minneapolis: University of Minnesota Press

Deleuze, G (1989) *Cinema II*. London: Continuum Press

Deleuze, G (2001) *Pure Immanence: Essays on a Life*. Cambridge, MA: MIT Press

Deleuze, G (2003 [1981]) *Francis Bacon: The Logic of Sensation*. London: Continuum Press

Deleuze, G (2004 [1968]) *Difference and Repetition*. London: Continuum Press

Deleuze, G (2006 [1986]) *Foucault*. London: Continuum Press

Deleuze, G (2007) *Two Regimes of Madness*. Los Angeles, CA: Semiotext(e) Publications

Deleuze, G; Guattari, F (1980) *Mille Plateaux*. Paris: Les Editions de minuit

Deleuze, G; Guattari, F (1986 [1975]) *Kafka: Towards a Minor Literature*. Minneapolis: University of Minnesota Press

Deleuze, G; Guattari, F (1994 [1991]) *What Is Philosophy?* London: Verso

Deleuze, G; Guattari, F (2004a [1984]) *Anti-oedipus*. London: Continuum Press

Deleuze, G; Guattari, F (2004b [1980]) *A Thousand Plateaus*. London: Continuum Press

Deleuze, G; Parnet, C (2002 [1977]) *Dialgues II*. London: Continuum Press

De Roo, G; Porter, G (2007) *Fuzzy Planning: The Role of Actors in a Fuzzy Governance Environment*. Aldershot: Ashgate

De Roo, G; Van Wezemael, J E; Hillier, J (Eds.) (2012) *Complexity and Planning: Systems, Assemblages and Simulations*. Surrey, UK: Ashgate

Design for London (2010) *London Housing Design Guide: Interim Edition*. London: London Development Agency

DfE (2011) Review of *Education Capital* by Sebastian James. London: UK Government Publications

Dovey, K (2013) 'Assembling architecture'. In Frichot, H; Loo, S (Eds.), *Deleuze and Architecture*. Edinburgh: Edinburgh University Press: 131–148

Dunn, W N (1994) *Public Policy Analysis: An Introduction* (2nd edn.). Englewood Cliffs, NJ: Prentice Hall

Dyke, S (2013) 'Disrupting "anorexia nervosa": An ethnography of the Deleuzian event'. In Coleman, R; Ringrose, J (Eds.), *Deleuze and Research Methodologies*. Edinburgh: Edinburgh University Press: 145–163

ESPON (2012) 'EATIA: ESPON and territorial impact assessment'. Final report: Version 29/06/2012

Evans, B (2007) 'Why we no longer need a town planning profession' *Planning Practice and Research* 8(1): 9–15

Fainstein, S S (2000) 'New directions in planning theory' *Urban Affairs Review* 35(4): 451–478

Fainstein, S S (2005) 'Planning theory and the city' *Journal of Planning Education and Research* 25(2): 121–130

Fancy, D (2011) 'Geoperformativity immanence, performance and the earth' *Performance Research* 16(4): 62–72

Farias, I; Bender, T (Eds.) (2010) *Urban Assemblages: How Actor-Network Theory Changes Urban Studies*. Oxon: Routledge

Faugier, J; Sargeant, M (1997) 'Sampling hard to reach populations' *Journal of Advanced Nursing* 26: 790–797

Fischer, T B (2001) *Strategic Environmental Assessment in Transport and Land Use Planning*. London, UK: Routledge

Fischer, T B (2003) 'Strategic environmental assessment in post-modern times' *Environmental Impact Assessment Review* 23(2): 155–170

Fischer, T B (2010) 'Reviewing the quality of strategic environmental assessment reports for English spatial plan core strategies' *Environmental Impact Assessment Review* 30: 62–69

Fischer, T B; Gazzola, P (2006) 'SEA effectiveness criteria – Equally valid in all countries? The case of Italy' *Environmental Impact Assessment Review* 26: 396–409

Fischer, T B; Seaton, K (2002) 'Strategic environmental assessment: Effective planning instrument or lost concept?' *Planning Practice and Research* 17(1): 31–44

Flathman, R B (1966) *The Public Interest: An Essay Concerning the Normative Discourse of Politics*. New York: Wiley

Flyvbjerg, B (1998) *Rationality and Power: Democracy in Practice*. Chicago: University of Chicago Press

Flyvjberg, B (2001) *Making Social Science Matter: Why Social Enquiry Fails and How It Can Suceed Again*. Cambridge, UK: Cambridge University Press

Flyvbjerg, B; Richardson, T (2002) 'Planning and Foucault: In search of the dark side of planning theory'. In Allmendinger, P; Tewdwr-Jones, M (Eds.), *Planning Futures: New Directions for Planning Theory*. London and New York: Routledge: 44–62

Food and Agriculture Organisation (2010) *Global Forest Resources Assessment, 2010*. Rome: FAO

Forester, J (2007) 'Reflections on trying to teach planning theory' *Planning Theory and Practice* 5(2): 242–251

Frichot, H; Gabrielsson, C; Metzger, J (2016) *Deleuze and the City*. Edinburgh: Edinburgh University Press

Frichot, H; Loo, S (Eds.) (2013) *Deleuze and Architecture*. Edinburgh: Edinburgh University Press

Friedmann, J (2008) 'The uses of planning theory: A bibliographic essay' *Journal of Planning Education and Research* 28: 247–257

Geertz, C (1973) *The Interpretation of Cultures*. New York: Basic Books

Gibbs, C; Gore, M L; McGarrell, E F; River, L (2010) 'Introducing conservation criminology: Towards interdisciplinary scholarship on environmental crimes and risks' *British Journal of Criminology* 50(1): 124–144

Giddens, A (1984) *The Constitution of Society: Outline of the Theory of Structuration*. Berkley, CA: University of California Press

Glynos, J; Howarth, D (2007) *Logics of Critical Explanation in Social and Political Theory*. Oxon: Routledge

Godoy, R; Groff, S; O'Neill, K (1998) 'The role of education in neotropical deforestation: Household evidence from Amerindians in Honduras' *Human Ecology* 26(4): 649–675

Godoy, R; Jacobson, M; De Castro, J; Aliaga, V; Romero, J; Davis, A (1998a) 'The role of tenure security and private time preferences in neotropical deforestation' *Land Economics* 74(2): 162–170

Gough, N (2008) 'Ecology, ecocriticism and learning: How do places *become* "pedagogical"?' *Transnational Curriculum Inquiry* 5(1) Available online: http://nitinat.library.ubc.ca/ojs/index.php/tci [Last accessed: January 2013]

Gower Davies, J (1972) *The Evangelistic Bureaucrat*. London: Tavistock

Graham, S; Healey, P (1999) 'Relational concepts of space and place: Issues for planning theory and practice' *European Planning Studies* 7(5): 623–646

Grinberg, S M (2013) 'Researching the pedagogical apparatus (dispositif): An ethnography of the molar, molecular and desire in contexts of extreme urban poverty'. In Coleman, R; Ringrose, J (Eds.), *Deleuze and Research Methodologies*. Edinburgh: Edinburgh University Press: 201–218

The Guardian (2012) 'Fears over green building standard for new schools' Thursday 26 January 2012. Available online: http://www.theguardian.com/environment/2012/jan/26/green-building-standard-new-schools [Last accessed: 18/03/2014]

Guattari, F (1984 [1977]) *Molecular Revolution: Psychiatry and Politics*. New York: Penguin

Guba, E G; Lincoln, Y S (1989) *Fourth Generation Evaluation*. Newbury Park, CA: Sage Publications

Guba, E G; Lincoln, Y S (1994) 'Competing paradigms in qualitative research'. In Denzin, N K; Lincoln, Y S (Eds.), *Handbook of Qualitative Research*. London: Sage Publications: 105–177

Gullison, R E; Losos, E C (1993) 'The role of foreign debt in deforestation in Latin America' *Conservation Biology* 7(1): 140–147

Gunder, M; Hillier, J (2004) 'Conforming to the expectations of the profession: A Lacanian perspective on planning practice, norms and values' *Planning Theory and Practice* 5(2): 217–235

Gunder, M; Hillier, J (2007) 'Problematising responsibility in planning theory and practice: On seeing the middle of the string' *Progress in Planning* 68(2): 57–96

Gunder, M; Hillier, J (2009) *Planning in Ten Words Or Less: A Lacanian Entanglement with Spatial Planning*. London: Ashgate

Hajer, M (1993) 'Discourse coalitions and the institutionalisation of practice. The case of acid rain in Britain.' In Forester, J; Fischer, F (Eds.), *The Argumentative Turn in Policy and Planning*. Durham: Duke University Press: 43–76

Hajer, M; Versteeg, W (2005) 'A decade of discourse analysis of environmental politics: Achievements, challenges, perspectives' *Journal of Environmental Policy & Planning* 7(3): 175–184

Hall, P; Pain, K (2006) *The Polycentric Metropolis: A Western European Perspective on Mega-city Regions*. Oxon: Earthscan publications

Halsey, M (2006) *Deleuze and Environmental Damage*. Aldershot, UK: Ashgate

Hammersley, M; Atkinson, P (2007) *Ethnography: Principles in Practice* (3rd edn.). London and New York: Routledge

Hamnett, C (2003) 'Contemporary human geography: Fiddling with Rome burns?' *Geofomm* 34(1): 1–3

Harper, T L; Stein, S M (2006) *Dialogical Planning in a Fragmented Society: Critically Liberal, Pragmatic, Incremental*. New Brunswick, NJ: CUPR Press

Harrison, P (2014) 'Making planning theory real' *Planning Theory* 13(1): 65–81

Hasson, F; Keeney, S (2011) 'Enhancing rigour in the Delphi technique research' *Technological Forecasting and Social Change* 78: 1695–1704

Healey, P (1996) 'The communicative turn in planning theory and its implications for spatial strategy formations' *Environment and Planning B: Planning and Design* 23(2): 217–234

Healey, P (1999) 'Institutionalist analysis, communicative planning, and shaping places' *Journal of Planning Education and Research* 19(2): 111–121

Healey, P (2003) 'Collaborative planning in perspective' *Planning Theory* 2(2): 101–123

Healey, P (2004a) 'Creativity and urban governance' *DISP* 158: 11–20

Healey, P (2004b) 'The treatment of space and place in the new strategic spatial planning in Europe' *International Journal of Urban and Regional Research* 28(1): 45–67

Healey, P (2006) 'Relational complexity and the imaginative power of strategic spatial planning' *European Planning Studies* 14(4): 525–546

Healey, P (2007) *Urban Complexity and Spatial Strategies: Towards a Relational Planning for Our Times*. New York: Routledge

Healey, P (2009a) 'The pragmatic tradition in planning thought' *Journal of Planning Education and Research* 28(3): 277–292

Healey, P (2009b) 'In search of the "strategic" in spatial strategy making' *Planning Theory & Practice* 10(4): 439–457

Healey, P (2010) *Making Better Places: The Planning Project in the Twenty-First Century*. New York: Palgrave Macmillan

Healey, P (2012) 'The universal and the contingent: Some reflections on the transnational flow of planning ideas and practices' *Planning Theory* 11(2): 188–207

Healey, P; Upton, R (Eds.) (2010) *Crossing Borders: International Exchange and Planning Practices*. London: Routledge

Heckathorn, D D (1997) 'Respondent-driven sampling: A new approach to the study of hidden populations' *Social Problems* 44: 174–199

Henkel, M (1991) *Government, Evaluation and Change*. London: Jessica Kingsley

Hillier, J (2003) 'Agon'izing over consensus: Why Habermasian ideals cannot be "real"' *Planning Theory* 2(1): 37–59

Hillier, J (2005) *Consent and Consensus*, edited collection with Cryle, D. Perth: API Network

Hillier, J (2007) *Stretching Beyond the Horizon: A Multiplanar Theory of Spatial Planning and Governance*. Aldershot: Ashgate

Hillier, J (2011) 'Strategic navigation across multiple planes: Towards a Deleuzian-inspired methodology for spatial planning' *Town Planning Review* 82(5): 503–527

Hillier, J; Abrahams, G (2013) *Deleuze and Guattari: Jean Hillier in Conversation with Gareth Abrahams*. AESOP InPlanning E-publication. DOI: 10.17418/B.2015.9789491937019

Hillier, J; Van Wezemael, J E (2012) 'On the emergence of agency in participatory strategic planning'. In De Roo, G; Van Wezemael, J E; Hillier, J (2012) (Eds.), *Complexity and Planning: Systems, Assemblages and Simulations*. Surrey, UK: Ashgate: 311–332

Howe, E (1994) *Acting on Ethics in City Planning*. New Brunswick: Rutgers, State University of New Jersey

HSE (no date) 'Designers: (Design and management) regulations 2007'. Guidance. Available online: http://www.hse.gov.uk/construction/cdm/faq/designers.htm [Last accessed: 29/07/2013]

HSE (1994) *Managing Health and Safety in Construction: Construction (Design and Management) Regulations 1994. Approved Code of Practice*. Sudbury: HSE Books

HSE (2007) *Managing Health and Safety in Construction: Construction (Design and Management) Regulations 2007. Approved Code of Practice*. Sudbury: HSE Books

HSE (2015) *Managing Health and Safety in Construction: Construction (Design and Management) Regulations 2015. Guidance on Regulations*. Sudbury: HSE Books

Hussler, C; Muller, P; Rondé, P (2011) 'Is diversity in Delphi panelist groups useful? Evidence from a French forecasting exercise on the future of nuclear energy' *Technological Forecasting and Social Change* 78: 1642–1653

Huxley, M (2000) 'The limits to communicative planning' *Journal of Planning Education and Research* 19(4): 369–377

ICOMOS (2010) *Guidance on Heritage Impact Assessments for Cultural World Heritage Properties*. Paris: ICOMOS Publications

ICS (2012) 'ICS news: What next for CDM? CDM 2014?'. Available online: http://www.instcs.org/news/what-next-for-cdm-2014.php [Last accessed: 22/07/2013]

Innes, J; Booher, D (1999) 'Consensus building and complex adaptive systems: A framework for evaluating collaborative planning' *Journal of the American Planning Association* 65(4): 412–423

Innes, J; Booher, D (1999a) 'Consensus building as role playing and bricolage: Toward a theory of collaborative' *Journal of the American Planning Association* 65(1): 9–26

Innes, J; Booher, D (2010) *Planning with Complexity: An Introduction to Collaborative Rationality for Public Policy*. London: Routledge

ISO (2011) *ISO 14006: 2011: Implementing and Managing Eco Design*. Geneva, Switzerland: ISO Publications

Jamal, T B; Stein, S M; Harper, T L (2002) 'Beyond labels: Pragmatic planning in multistakeholder tourism–environmental conflicts' *Journal of Planning Education and Research* 22(2): 164–177

Jha, S; Bawa, K S (1999) 'Population growth, human development, and deforestation in biodiversity hotspots' *Conservation Biology* 20(3): 906–912

Jones, M (2010) 'Phase space: Geography, relational thinking, and beyond' *Progress in Human Geography* 33(4): 487–506

Kaufman, E (1998) 'Introduction'. In Kaufman, E; Heller, K J (Eds.), *Deleuze and Guattari: New Mappings in Politics, Philosophy and Culture*. Minneapolis, MN: University of Minnesota Press: 3–19

Khakee, A (1998) 'Evaluation and planning: Inseparable concepts' *The Town Planning Review* 69(4): 359–374

Khakee, A (2003) 'The emerging gap between evaluation research and practice' *Evaluation* 9(3): 340–352

Khakee, A; Barbanente, A; Borri, D (2000) 'Expert and experiential knowledge in planning' *Journal of the Operational Research Society* 51(7): 776–788

Kitzinger, J (1995) 'Qualitative research: Introducing focus groups' *British Medical Journal* 311: 299–302

Klosterman, R E (1980) 'A public interest criterion' *Journal of the American Planning Association* 46(3): 323–333

Klosterman, R E (1985) 'Arguments for and against planning' *Town Planning Review* 56(1): 5–20

Laclau, E (1996) *Emancipation(s)*. London: Verso

Laclau, E (2003) 'Why do empty signifiers matter to politicians?' In Zizek, S (Ed.), *Jacques Lacan, Volume III*. London: Routledge: 305–311

Laclau, E (2005) *On Populist Reason*. London: Verso

Law J (2004) 'And if the global were small and noncoherent? Method, complexity, and the baroque' *Environment and Planning D: Society and Space* 22(1): 13–26

Lawrence, D P (2000) 'Planning theories and environmental impact assessment' *Environmental Impact Assessment Review* 20(6): 607–625

Lawson, J (2006) *Critical Realism and Housing Research*. Oxon: Routledge

Lichfield, N (2001) 'The philosophy and role of community impact evaluation in the planning system'. In Voogd, H (Ed.), *Recent Developments in Evaluation*. Groningen, Netherlands: Geopress: 153–173

Lichfield, N; Prat, A (1998) 'Linking ex ante and ex post evaluation in British town planning'. In Lichfield, N (Ed.), *Evaluation in Planning: Facing the Challenge of Complexity*. Dordrecht, Netherlands: Kluwer Academic Publishers: 283–298

Lincoln, Y S; Denzin, N K (2003) *Turning Points in Qualitative Research: Tying Knots in a Handkerchief.* Walnut Creek, CA: AltaMira Press

Lincoln, Y S; Lynham, S A; Guba, E G (2011) 'Paradigmatic controversies, contradictions, and emerging confluences revisited'. In Denzin, N K; Lincoln, Y S (Eds.), *The Sage Handbook of Qualitative Research* (4th Ed.). London: Sage Publications: 97–128

Linstone, H A; Turoff, M (1975) *The Delphi Method: Techniques and Applications*. Boston: Addison-Wesley Educational Publishers

Lofstedt, R E (2011) 'Reclaiming health and safety for all: An independent review of health and safety legislation'. London: Department for Work and Pensions Available online: https://www.gov.uk/government/uploads/system/uploads/attachment_data/file/66790/lofstedt-report.pdf. [Last accessed: 08/05/2014]

Lord, A (2014) 'Towards a non-theoretical understanding of planning' *Planning Theory* 13(1): 26–43

Lowri, K (1994) 'The legitimation of planning' *Planning Theory* 6(10–11): 99–109

Lynch, M; Stretsky, P (2003) 'The meaning of green: Contrasting criminological perspectives' *Theoretical Criminology* 7(2): 217–238

Maginn, P J (2007) 'Towards more effective community participation? The potential of collaborative planning and applied ethnography' *Qualitative Research* 7(1): 25–43

Malpas, J (2012) 'Putting space in place: Philosophical topography and relational' Geography' *Environment and Planning D: Society and Space* 30(2): 226–242

March, A (2010) 'Practising theory: When theory affects urban planning' *Planning Theory* 9(2): 1108–1125

Markey, S; Connelly, S; Roseland, M (2010) '"Back of the envelope": Pragmatic planning for sustainable rural community development' *Planning, Practice & Research* 25(1): 1–23

Marston, S A; Jones, J P; Woodward, K (2005) 'Human geography without scale' *Transactions of the Institute of British Geographers* 30(4): 416–432

Martin, R (2001) 'Geography and public policy: The case of the missing agenda' *Progress in Human Geography* 25(2): 189–210

Martin-Jones, D; Brown, W (Eds.) (2012) *Deleuze and Film*. Edinburgh: Edinburgh University Press

Massey, D (1999) 'Space-time, "science" and the relationship between physical geography and human geography' *Transactions of the Institute of British Geographers* 24(3): 261–276

Massumi, B (2004 [1980]) 'Translator's foreword: Pleasures of philosophy'. In Deleuze, G; Guattari, F (Eds.), *A Thousand Plateaus*. London: Continuum Press: ix–xvi

Massumi, B (2010) 'What concepts do: Preface to the Chinese translation of *A Thousand Plateaus*' *Deleuze Studies* 4: 1–15

Massumi, B (2013) *Semblance and Event: Activist Philosophy and the Occurrent Arts*. Cambridge, MA: MIT Press

Mead D; Mosley L (2001) 'The use of Delphi as a research approach' *Nurse Researcher* 8: 4–23

Meyerson, M; Baufald, E (1955) *Politics, Planning and the Public Interest*. New York: The Free Press

Mikulak, M (2007) 'Cross-pollinating marxism and deep ecology: Towards a post-humanist eco-humanism' *Cultural Logic: Marxism Theory and Practice*. Available online: http://clogic.eserver.org/2007/2007.html [Last accessed: June 2013]

Miller, J; Glassner, B (2010) 'The "inside" and the "outside": Finding realities in interviews'. In Silverman, D (Ed.), *Qualitative Research*. Thousand Oaks, CA: Sage Publications: 132

Mitchell, J C (1983) 'Case and situation analysis' *Sociological Review* 31(2): 187–211

Monno, V (2012) 'A different view of relational complexity: Imagining places through the Deleuzean social cartography.' In De Roo, G; Hillier, J; Van Wezemael, J (Eds.), *Complexity and Planning: Systems, Assemblages and Simulations*. Surrey, UK: Ashgate: 287–310

Monroe, D K; Müller, D (2007) 'Issues in spatially explicit statistical land-use/cover change (LUCC) models: Examples from western Honduras and the Central Highlands of Vietnam' *Land Use Policy* 24(3): 521–530

Moroni, S (2004) 'Towards a reconstruction of the public interest criterion' *Planning Theory* 3(2): 151–171

Moroni, S (2010) 'Rethinking the theory and practice of land-use regulation: Towards nomocracy' *Planning Theory* 9(2): 137–155

Murdoch, J (1995) 'Actor-networks and the evolution of economic forms: Combining description and explanation in theories of regulation, flexible specialization, and networks' *Environment and Planning A* 27(5): 731–757

Murdoch, J (1997) 'Inhuman/nonhuman/human: Actor-network theory and the prospects for a nondualistic and symmetrical perspective on nature and society' *Environment and Planning D: Society and Space* 15(6): 731–756

Murdoch, J (1998) 'The spaces of actor-network theory' *Geoforum* 29(4): 357–374

Murdoch, J (2001) 'Actor-network theory, co-construction and the problem of human exemptionalism' *Sociology* 35(1): 111–133

Murdoch, J (2006) *Post-structuralist Geography: A Guide to Relational Space*. London: Sage Publications

Nagendra, H; Monroe, D K; Southworth, J (2004) 'From pattern to process: Landscape fragmentation and the analysis of land use/land cover change' *Agriculture, Ecosystems & Environment* 101(2–3): 111–115

The National Affordable Housing Agency (2007) '721 Housing Quality Indicators (HQI) For'. Version 4. Published May 2007. Updated April 2008. Housing Corporation

Neuman, M (2005) 'Notes on the uses and scope of city planning theory' *Planning Theory* 4(2): 123–145

Normark, J (2009) 'The making of a home: Assembling houses at Nohcacab, Mexico' *World Archaeology* 41(3): 430–444

Nowack, M; Endrikat, J; Geunther, E (2011) 'Review of Delphi-based scenario studies: Quality and design considerations' *Technological Forecasting and Social Change* 78: 1603–1615

Nyseth, T; Pløger, J; Holm, T (2009) 'Planning beyond the horizon: The Tromsø experiment' *Planning Theory* 9(3) 223–247

OECD (2009) 'Sustainable development: Linking economy, society, environment'. OECD Publications

Office for National Statistics (2011a) *Measuring Economic Well-being*. London: Office for National Statistics Publications

Office for National Statistics (2011b) *Measuring What Matters: National Statistician's Reflections on the National Debate on Measuring National Well-being*. London: Office for National Statistics Publications

Owens, S (1994) 'Land, limits and sustainability: A conceptual framework and some dilemmas for the planning system' *Transactions* 19: 439–456

Palmas, K (2008) 'DDR, 1989: A Deleuze-inspired account'. Paper presented at the Political Theory Seminar, the Department of Political Science, Göteborg University, 30 January 2008

Parr, A (Ed.) (2005) *The Deleuze Dictionary*. Edinburgh, UK: Edinburgh University Press

Patton, P (2000) *Deleuze and the Political*. London: Routledge

Patton, P (2010) *Deleuzian Concepts: Philosophy, Colonization, Politics*. Stanford: Stanford University Press

Pauleit, S; Ennos, R; Golding, Y (2005) 'Modeling the environmental impacts of urban land use and land cover change – A study in Merseyside, UK' *Landscape and Urban Planning* 71(2–4): 295–310

Pfeffer, M J; Schlelhas, J S; DeGloria, S D; Gomez, J (2005) 'Population, conservation, and land use change in Honduras' *Agriculture, Ecosystems & Environment* 101(1–2): 14–28

Pierre, N J (1999) 'Models of urban governance: The institutional dimension of urban politics' *Urban Affairs Review*, 34(3): 372–396

Pine, A (2008) *Working Hard, Drinking Hard: On Violence and Survival in Honduras*. Oakland, CA: University of California Press

Pløger, J (2010a) 'Contested urbanism: Struggles about representations' *Space and Polity* 14(2): 143–165

Pløger, J (2010b) 'Urbanity, (neo)vitalism and becoming'. In Hillier, J; Healey, P (Eds.), *Conceptual Challenges for Planning Theory*. Aldershot: Ashgate: 319–342

Porter, L; Davoudi, S (2012) 'The politics of resilience for planning: A cautionary note' *Planning Theory & Practice* 13(2): 329–333

Poulton, M C (1991) 'The case for a positive theory of planning. Part 1: What is wrong with planning theory?' *Environment and Planning B: Planning and Design* 18: 225–232

Purcell, M (2013) 'A new land: Deleuze and Guattari and planning' *Planning Theory* 14(1): 20–38

Reade, E (1997) 'Planning in the future or planning of the future'. In Blowers, A; Evans, B (Eds.), *Town Planning into the 21st Century*. London: Routledge: 71–103

Reed. M I (2009) 'Critical realism: Philosophy, method, or philosophy in search of a method'? In Buchanan, A; Bryman, A (Eds.), *The Sage Handbook of Organisational Research Methods*. London: Sage Publications: 430–448

Renold, E; Mellor, D (2013) 'Deleuze and Guattari in the nursery: Towards an ethnographic multi-sensory mapping of gendered bodies and becomings'. In Coleman, R; Ringrose, J (Eds.), *Deleuze and Research Methodologies*. Edinburgh: Edinburgh University Press: 23–41

Richardson, T (2005) 'Environmental assessment and planning theory: Four short stories about power, multiple rationality, and ethics' *Environmental Impact Assessment Review* 25(4): 341–365

Richardson, T; Jensen, O B (2001) 'Nested visions: New rationalities of space in European spatial planning' *Regional Studies* 35(8): 703–717

Richardson, T; Jensen, O B (2003) 'Linking discourse and space: Towards a cultural sociology of space in analysing spatial policy discourses' *Urban Studies* 40(1): 7–22

Robbins, P; Marks, B (2010) 'Assembling geographies'. In Smith, S J; Pain, R; Marston, S A; Jones, J P (Eds.), *The SAGE Handbook of Social Geographies*. London: Sage Publications: 176–194

Roller, L H; Waverman, L (2001) 'Telecommunications infrastructure and economic development: A simultaneous approach' *American Economic Review* 91(4): 909–923

Rorty, R (1982) *Consequences of Pragmatism*. Minneapolis: University of Minnesota Press

Rowe, G; Wright, G (1999) 'The Delphi technique as a forecasting tool: Issues and analysis' *International Journal of Forecasting* 15: 353–375

Rowe, G; Wright, G (2011) 'The Delphi technique: Past, present, and future prospects – Introduction to the special issue' *Technological Forecasting and Social Change* 78: 1487–1490

Rowe, G; Wright, G; McColl, A (2005) 'Judgment change during Delphi-like procedures: The role of majority influence, expertise, and confidence' *Technological Forecasting and Social Change* 72(4): 377–399

Ruf, F; Schroth, G (2004) 'Chocolate forests and monocultures: A historical review of cocoa growing and its conflicting role in tropical deforestation and forest conservation'. In Schroth, G; da Fonseca, A B; Harvey, C A; Gascon, C; Vasconcelos, H L; Izac, A N (Eds.), *Agroforestry and Biodiversity Conservation in Tropical Landscapes*. Washington, DC: Island Press: 107–134

Rydin, Y (2014) 'The challenges of the "material turn" for planning studies' *Planning Theory and Practice* 15(4): 590–595

Sawyer, L; Wilde, P D; Turpin-Brooks, S (2008) 'Energy performance and occupancy satisfaction: A comparison of two closely related buildings' *Facilities* 26(13/14): 542–551

Sayer, A (1992) *Method in Social Science*. London: Routledge

Sayer, A (1997) 'Essentialism, social constructionism, and beyond' *The Sociological Review* 45(3): 453–487

Sayer, A (2000) *Realism and Social Science*. London: Sage Publications

Schweber, L (2013) 'The effect of BREEAM on clients and construction professionals' *Building Research and Information* 41(2): 1–17

Scott, A J; Roweis, S T (1977) 'Urban planning and theory and practice: A reappraisal' *Environment and Planning A* 9(10): 1097–1119

Searle, G; Bunker, R (2010) 'Metropolitan strategic planning: An Australian paradigm?' *Planning Theory* 9(3) 163–180

Smith, D W (2006) 'The concept of the simulacrum: Deleuze and the overturning of Platonism' *Continental Philosophy Review* 38: 89–123

Söderbaum, P (1998) 'Economics and ecological sustainability: An actor-network approach to evaluation'. In Lichfield, N; Barbanente, A; Borri, D; Khakee, A; Prat, A (Eds.), *Evaluation in Planning: Facing the Challenge of Complexity*. Dordrecht: Kluwer: 51–72

Sridhar, K S; Sridhar, V (2007) 'Telecommunications infrastructure and economic growth: Evidence from developing countries' *Applied Econometrics and International Development* 7(2): 37–56

Stein, S M; Harper, T L (2012) 'Creativity and innovation: Divergence and convergence in pragmatic dialogical planning' *Journal of Planning Education and Research* 32(1): 5–17

Tait, M (2011) 'Trust and the public interest in the micropolitics of planning practice' *Journal of Planning Education and Research* 31(2): 157–171

Tait, M; Jensen, O (2007) 'Travelling ideas, power and place: The cases of urban villages and business improvement districts' *International Planning Studies* 12(2): 107–128**Thompson, R (2010)** 'Re-defining planning: The roles of theory and practice' *Planning Theory and Practice* 1(1): 126–133

Thrift, N (2004) 'Remembering the technological unconscious by foregrounding knowledges of position' *Environment and Planning D: Society and Space* 22(1): 175–190

Thrift, N (2007) *Non-representational Theory: Space, Politics, Affect*. Oxon, UK: Routledge

Trummer, P (2012) 'Population thinking in architecture'. In De Roo, G; Van Wezemael, J E; Hillier, J (Eds.), *Complexity and Planning: Systems, Assemblages and Simulations*. Surrey, UK: Ashgate: 333–346

UK Government (2011) 'Building regulations and related legislation'. Available online: http://www.redtapechallenge.cabinetoffice.gov.uk/building-regulations-and-related-legislation/ [Last accessed: 04/09/2013]

UNESCO (2013) *Operational Guidelines for the Implementation of the World Heritage Convention*. Paris: UNESCO World Heritage Centre

Van Wezemael, J E (2008) 'The contribution of assemblage theory and minor politics for democratic network governance' *Planning Theory* 7(2): 165–185

Van Wezemael, J E (2010) 'Modulation of singularities – A complexity approach to planning competitions'. In Hillier, J; Healey, P (Eds.), *Conceptual Challenges for Planning Theory*. Aldershot: Ashgate: 273–290

Van Wezemael, J E (2012) 'Transformative practice as an exploration of possible spaces'. In De Roo, G; Van Wezemael, J E; Hillier, J (Eds.), *Complexity and Planning: Systems, Assemblages and Simulations*. Surrey, UK: Ashgate: 93–113

Verhage, R (2003) 'The role of the public sector in urban development: Lessons from Leidsche Rijn Utrecht (The Netherlands)' *Planning Theory and Practice* 4(1): 29–44

Vigar, G (2012) 'Planning and professionalism: Knowledge, judgement and expertise in English planning' *Planning Theory* 11(4): 361–378

Waterhout, B (2007) 'Territorial cohesion: The underlying discourses'. In Faludi, A (Ed.), *Territorial Cohesion and the European Model of Society*. Cambridge, MA: Lincoln Institute of Land Policy: 37–59

Wells, J W (2007) 'Enemies of the roof'. Available online: http://www.rci-online.org/interface/2007-03-wells.pdf. [Last accessed: 05/05/2014]

Whittle, A; Spicer, A (2008) 'Is actor-network theory critique?' *Organization Studies* 29(4): 611–629

Wilkinson, C (2011) 'Strategic navigation: In search of an adaptable mode of strategic spatial planning practice' *Town Planning Review* 82(5): 595–613

Wilkinson, C (2012) 'Social-ecological resilience: Insights and issues for planning theory' *Planning Theory* 11(2): 148–169

Williams, J (2008) *Gilles Deleuze's Logic of Sense: A Critical Introduction and Guide*. Edinburgh: Edinburgh University Press

Willmott Dixon (2010) 'Building Research Establishment Environmental Assessment Method (BREEAM)'. Briefing Note 16. Available online: http://www.willmottdixon group.co.uk/assets/b/r/briefing-note-16-breeam-2.pdf [Last accessed: 19/03/2014]

Wood, S (2009) 'Desiring docklands: Deleuze and urban planning discourse' *Planning Theory* 8(2): 191–216

Wright, G; Rowe, G (2011) 'Group-based judgmental forecasting: An integration of extant knowledge and the development of priorities for a new research agenda' *International Journal of Forecasting* 27: 1–13

Yeung, H W (1997) 'Critical realism and realist research in human geography: A method or a philosophy in search of a method?' *Progress in Human Geography* 21(1): 51–57

Yiftachel, O (1989) 'Towards a new typology of urban planning theories' *Environment and Planning B: Planning and Design* 16: 23–39

Yin, R K (2009) *Case Study Research: Design and Methods* (4th edn.). Thousand Oaks, CA: Sage Publications

Index

Note; Page numbers in italics indicate figures and tables.

For Product Safety Concerns and Information please contact our EU
representative GPSR@taylorandfrancis.com
Taylor & Francis Verlag GmbH, Kaufingerstraße 24, 80331 München, Germany

www.ingramcontent.com/pod-product-compliance
Ingram Content Group UK Ltd.
Pitfield, Milton Keynes, MK11 3LW, UK
UKHW022129190926
475702UK00013B/1703
* 9 7 8 1 1 3 8 3 9 2 8 0 9 *